Water Supply System: Theory, Analytical Aspects and Models

Water Supply System: Theory, Analytical Aspects and Models

Edited by **Danice Coy**

New York

Published by Callisto Reference,
106 Park Avenue, Suite 200,
New York, NY 10016, USA
www.callistoreference.com

Water Supply System: Theory, Analytical Aspects and Models
Edited by Danice Coy

International Standard Book Number: 978-1-63239-621-1 (Hardback)

Contents

Preface

This book was inspired by the evolution of our times; to answer the curiosity of inquisitive minds. Many developments have occurred across the globe in the recent past which has transformed the progress in the field.

This book is a comprehensive study of modern and traditional water supply systems. A water supply system is an interrelated compilation of sources, pipes, and hydraulic control elements delivering set water quantities at preferred pressures and water qualities. This book incorporates chosen topics on theory, review and realistic function models for water supply systems study, including, strategy for temporary analysis, sustainable management of local water supply systems, infrastructure asset administration, optimal pump preparation, demand improbability, errors in water meter measuring and indicators for water mains treatment.

This book was developed from a mere concept to drafts to chapters and finally compiled together as a complete text to benefit the readers across all nations. To ensure the quality of the content we instilled two significant steps in our procedure. The first was to appoint an editorial team that would verify the data and statistics provided in the book and also select the most appropriate and valuable contributions from the plentiful contributions we received from authors worldwide. The next step was to appoint an expert of the topic as the Editor-in-Chief, who would head the project and finally make the necessary amendments and modifications to make the text reader-friendly. I was then commissioned to examine all the material to present the topics in the most comprehensible and productive format.

I would like to take this opportunity to thank all the contributing authors who were supportive enough to contribute their time and knowledge to this project. I also wish to convey my regards to my family who have been extremely supportive during the entire project.

Editor

Infrastructure Asset Management of Urban Water Systems

Helena Alegre and Sérgio T. Coelho

Additional information is available at the end of the chapter

1. Introduction

Urban water systems are the most valuable part of the public infrastructure worldwide, and utilities and municipalities are entrusted with the responsibility of managing and expanding them for current and future generations. Infrastructures inexorably age and degrade, while society places increasing demands for levels of service, risk management and sustainability.

As many systems reach high levels of deferred maintenance and rehabilitation (ASCE, 2009), the combined replacement value of such infrastructures is overwhelming, demanding judicious spending and efficient planning.

Infrastructure asset management (IAM) of urban water infrastructures is the set of processes that utilities need to have in place in order to ensure that infrastructure performance corresponds to service targets over time, that risks are adequately managed, and that the corresponding costs, in a lifetime cost perspective, are as low as possible.

IAM methods partially differ from those applicable to managing other types of assets. One of the reasons is the fact that such infrastructures have indefinite lives, in order to satisfy the permanent needs of a specific public service. Infrastructures are not replaceable as a whole, only piecemeal. Consequently, in a mature infrastructure, all phases of assets lifetime coexist. Additionally, in network-based infrastructures, it is frequently not feasible to allocate levels of service to individual components because there is a dominant system behavior (e.g. symptoms and their causes often occur at different locations).

IAM is increasingly becoming a key topic in the move towards compliance with performance requirements in water supply and wastewater systems. Sustainable management of these systems should respond to the need for:

- Promoting adequate levels of service and strengthening long-term service reliability;
- Improving the sustainable use of water and energy;
- Managing service risk, taking into account users' needs and risk acceptance;
- Extending service life of existing assets instead of building new, when feasible;
- Upholding and phasing in climate change adaptations;
- Improving investment and operational efficiency in the organization;
- Justifying investment priorities in a clear, straightforward and accountable manner.

2. Overview of current knowledge and practice

Given its origin in the financial sector, where the economic approach is prevalent, the first significant developments in the field of infrastructure asset management were led by accountants and economists. In the late 1980s, the South Australia Public Accountants Committee published a series of eight reports alerting all Australian governments for the need to seriously consider the management of their infrastructure if deterioration of valuable public services were to be avoided (Burns et al., 1999). Following these reports, Prof. Penny Burns, at the University of Adelaide (Australia), played an crucial role in bringing to attention the importance of the subject and formalizing key concepts and principles (e.g., Burns, 1990; Burns et al., 1999). Australian leadership in this field endures to the present day, through both industry practice and initiatives by organizations such as the Institute of Public Works Engineering Australia (IPWEA, www.ipea.org.au), the National Asset Management Steering Group (NAMS, www.nams.au.com), the Australian National Audit Office (ANAO, www.anao.gov.au), the Asset Management Quarterly International (AMQI, www.amqi.com), ACORN Inc. (www.acorninc.org) and the Water Services Association Australia (WSAA, www.wsaa.asn.au).

The Australian and New Zealand AM school is synthesized in the International Infrastructure Management Manual, revised and updated periodically (current edition: IIMM, 2011), which addresses different types of public infrastructures and promotes the Total Asset Management Process.

IAM has equally seen significant advances in many other countries, such as in the US (e.g. Clark et al. 2010; US EPA, 2012), the UK (e.g. IAM/BSI, 2008; UKWIR, 2003) and Portugal (Alegre and Covas, 2010; Coelho and Vitorino, 2011; Alegreet al., 2011). From a practical standpoint, very good examples of leading-edge utility practice can be found in Asia (e.g., Singapore PUB), and in Central and Northern Europe, such as in the Netherlands (e.g. PWN - North Holland), Germany (e.g. Munich, Berlin), Norway (e.g. Oslo) or Sweden (e.g. Stockholm, Malmo).

IAM has also registered scientific developments, particularly with regard to algorithms and tools aiming at supporting pipe rehabilitation prioritization and decision-making. Whole-life costing (e.g. Skipworthet al., 2002), as well as life time assessment and failure forecasting, are among the most researched topics (e.g., Sægrov ed., 2005; Sægrov ed., 2006; Malm*et al.*, 2012; Renaud *et al.*, 2011). From 2005, the biannual LESAM (Leading-Edge Strategic Asset Management) conferences of the International Water Association have clearly demonstrated the increasing interest and recognition of this field of knowledge (e.g. Alegre and Almeida ed., 2009).

Effective decision-making requires a comprehensive approach that ensures the desired performance at an acceptable risk level, taking into consideration the costs of building, operating, maintaining and disposing capital assets over their life cycles. Brown and Humphrey (2005) summarize these concepts by defining IAM as "the art of balancing performance, cost and risk in the long-term".

IAM is most often approached based on partial views: e.g., for business managers and accountants, IAM means financial planning and the control of business risk exposure (Harlow and Young, 2001); for water engineers, IAM is focused on network analysis and design, master planning, construction, optimal operation and hydraulic reliability (Alegre and Almeida ed., 2009); for asset maintenance managers, the infrastructure is mostly an inventory of individual assets and IAM tends to be the sum of asset-by-asset plans, established based on condition and criticality assessment; for many elected officials, since water infrastructures are mostly buried, low visibility assets, IAM tends to be driven by service coverage, quality and affordability in the short run. Common misconceptions include reducing IAM to a one-size-fits-all set of principles and solutions, mistaking it for a piece of software, substituting it for engineering technology, or believing that it can be altogether outsourced. In practical terms, many existing implementations tend to be biased by one or several of these perspectives.

3. IAM as an integrated approach

To avoid the shortcomings inherent to these partial views, integrated IAM approaches are required, driven by the need to provide adequate levels of service and a sustainable service in the long-term.

Integrated IAM may be implemented in many different forms. Even for a specific utility and a given external context, different approaches may be successfully implemented. However, there are some basic principles commonly accepted in the current leading literature, practice and standardization (Hughes, 2002; INGENIUM and IPWEA, 2011; Sægrov ed., 2005; Sægrov ed., 2006; Sneesby, 2010).

An integrated methodology is presented that approaches IAM as a management process, based on PDCA principles and requiring full alignment between the strategic objectives and targets, and the actual priorities and actions implemented, embedding the key requirements of the forthcoming ISO 55000/55001/55002 standards on asset management (ISO, 2012a, 2012b, 2012c). The approach expressly takes into account that a networked infrastructure cannot be dealt with in the same way as other collections of physical assets: it has a dominant system behavior (i.e., individual assets are not independent from one another), and as a whole it does not have a finite life – it cannot be replaced in its entirety, only piecemeal (Burns *et al.*, 1999). The methodology allows for the assessment and comparison of intervention alternatives from the performance, cost and risk perspectives over the analysis hori-zon(s), taking into account the objectives and targets defined (Alegre and Covas, 2010; Almeida and Cardoso, 2010). In summary, the objective of an integrated approach is to assist water utilities in answering the following questions:

• Who are we at present, and what service do we deliver?

• What do we own in terms of infrastructures?

• Where do we want to be in the long-term?

• How do we get there?

The cube shown in Fig. 1 symbolizes an integrated IAM approach. It advocates that IAM must be addressed at different planning decisional levels: a strategic level, driven by corporate and long-term views and aimed at establishing and communicating strategic priorities to staff and citizens; a tactical level, where the intermediate managers in charge of the infrastructures need to select what the best medium-term intervention solutions are; and an operational level, where the short-term actions are planned and implemented. It also draws attention to the need for standardized procedures to assess intervention alternatives in terms of performance, risk and cost, over the analysis period. The other relevant message is that IAM requires three main pillars of competence: business management, engineering and information.

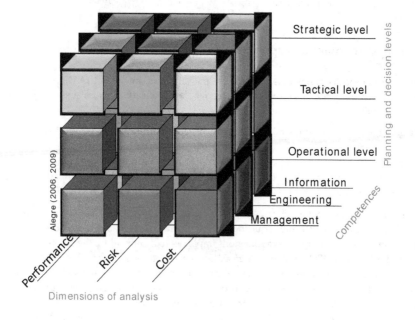

Figure 1. General IAM approach

At each level of management and planning –strategic, tactical and operational –a structured loop (Fig. 2) comprises the following stages: (i) definition of objectives and targets; (ii) diagnosis; (iii) plan production, including the identification, comparison and selection of alternative solutions; (iv) plan implementation; and (v) monitoring and review. Most utilities already have several elements of this process in place. What is often missing is a review mechanism – a way to measure compliance with set goals – as well as an effective alignment between the different management levels.

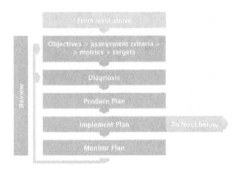

Figure 2. The planning process at each planning level

Setting up objectives, assessment criteria, metrics and targets is a crucial stage in order to set up clear directions of action, as well as accountability of results through timely review, within a given time frame (short, medium or long-term) (ISO 24510:2007, 24511:2007, 24512:2007). These metrics and targets are an essential basis for establishing the diagnosis, prioritizing intervention solutions and monitoring the results.

The process cascades through the decisional levels within the organization's management structure. The global approach is based on plan-do-check-act (PDCA) principles aiming at the continuous improvement of the IAM process. The key notions in this process are alignment among the decisional levels and their actors; bottom-up feedback; and involvement and empowerment of the entire organization, from the CEO to the asset operators, in order to promote leadership, co-ordination, collaboration, corporate culture acceptance, motivation, commitment and corporate know-how.

4. From whole-life costing to long-term analysis of indefinite life systems

Comparing intervention alternatives from the financialstandpointrequires that all relevant costs and revenues incurred during the asset life be taken into account. The costs in particular include such items as design and building costs, operating costs, maintenance costs, associated financing costs, depreciation, and disposal costs. Most of the reference literature on asset management recommends a whole-life costing approach (also known as life-cycle approach). However, this is not directly applicable to urban water infrastructures and other networked infrastructures that have indefinite lives and behave as systems, not as mere collections of components with independent functionality.

As argued by Burns *et al.* (1999), infrastructure assets are defined functionally as assets that are not replaced as a whole but rather are renewed piecemeal through the replacement of individual components, whilst maintaining the overall function of the system. As a whole, infrastructure system assets have indefinite lives. Conversely, economic lives can only be assigned to the individual components of an infrastructure system.

However, intervention decisions cannot be made based exclusively on the analysis of each individual asset. Individual assets cannot deliver a service by themselves, but only as part of a system or subsystem. The causes of malfunctions are often located away from where the symptoms emerge. Levels of service cannot be allocated to individual assets, for most of the infrastructure's components. Intervention alternatives, aimed at producing the desire deffect, tend to imply jointly modifying a combination of assets, which display different remaining lives, values, condition, etc..

These two key features – the indefinite life of the infrastructure as a whole,and its system behavior – make the classical life-cycle approach effectively unsuitable to IAM. The objective is to ensure that the service provided meets the targets over time, keeping the risk in acceptable levels and minimizing the overall costs from a long run viewpoint.

How long is "long-term"? Long enough that interventions are given time to reach their infrastructural maturity, all the lifecycle stages of the most relevant assetsare included in a meaningful way,and the investments under consideration are rewarded by their accrued benefits; but not so long into the future as to unreasonably limit the significance of the assumptions made for the scenarios considered, such as demand or land use projections.

5. Performance, risk and cost

5.1. Performance assessment

As previously mentioned, IAM aims at ensuring that, in a long-term perspective, service performance is kept adequate, risks incurred are acceptable and the corresponding costs are as low as feasible. Assessing performance, risk and cost is therefore key to effective IAM.

Performance may translate by either the efficiency or the effectiveness of the service. Performance assessment is a widespread activity used in economics, business, sports and many other walks of life in general, in order to compare and score entities and individuals and take management or other decisions (Alegre*et al.*, 2000, Matos *et al.*, 2003, Alegre*et al.* 2006, Cabrera &Pardo, 2008, Sjovold*et al.* eds., 2008, ISO 24510, ISO 24511, ISO 24512).

Assessment is defined as a "process, or result of this process, that compares a specified subject matter to relevant references" (ISO 24500).Performance assessment is therefore any approach that allows for the evaluation of the efficiency or the effectiveness of a process or activity through the production of performance measures. Performance measures are the specific parameters that are used to inform the assessment. The principal categories of performance measures include (Sjovold*et al.* eds., 2008):

- Performance indicators, which are quantitative efficiency or effectiveness measures for the activity of a utility. A performance indicator consists of a value (resulting from the evaluation of the "processing rule") expressed in specific units, and a confidence grade which indicates the quality of the data represented by the indicator. Performance Indicators are typically expressed as ratios between variables; these may be commensurate (e.g. %) or non-commensurate (e.g. $/m3). The information provided by a performance indicator is the result of a comparison (to a target value, previous values of the same indicator, or values of the same indicator from other undertakings) (Alegre*et al.* 2006; ISO 24500, Sjovold*et al.* eds., 2008).

- Performance indices, which are standardised and commensurable measures, may result from the combination of more disaggregated performance measures (e.g. weighted average of performance indicators) or from analysis tools (e.g. simulation models, statistical tools, cost efficiency methods). Sometimes they aim at aggregating several perspectives into in a single measure (Alegre, 2008, Sjovold*et al.* eds., 2008).Differently from the performance indicators, they contain a judgment in itself, intrinsic to the standardization process (e.g. 0 – no function; 1 – minimum acceptable; 2 – good; 3 – excellent).

- Performance levels, which are performance measures of a qualitative nature, expressed in discrete categories (e.g. excellent, good, fair, poor). In general they are adopted when the use of quantitative measures is not appropriate (e.g. evaluation of customer satisfaction by means of surveys) (Alegre, 2008, Sjovold*et al.* eds., 2008).

Performance indicators may be converted into performance indices through the application of a performance function, or into performance levels when they are compared with reference levels, in order to support interpretation or multi-criteria analyses. Such transformations may be particularly useful in the graphical representation of a set of performance indicators.

5.2. Risk assessment

Risk analysis may address an organization in its entirety, a system or sub-systems (aggregated or lumped analysis), or individual system components(component or discrete analysis). Risk assessment may be carried outin many different ways, and is often (though not always) quantifiable: for instance, if the probability of failure of every pipe in a network is known, as well as its consequence, expressed in terms of the ensuing reduced service (unmet demand), the total risk of not supplying the users may be expressed as the expected value of the annual unmet demand (Vitorino*et al.*, 2012).

Risk analysis is a vast field of expertise where several mainstream frameworks have been developed for infrastructure-based problems, such as fault-tree analysis or the approaches centered on risk matrices (Almeida *et al.*, 2010). The latter is one of the most versatile and structured formalisms available when approaching the range of (quantifiable or unquantifiable) risks that are faced by urban utilities, and is based on a thorough analysis of risk consequences and on the categorization into both probability and consequenceclasses.

Probability classes can be defined by different probability intervals that may be derived, typically, from linear, exponential or logarithmic functions. The selection of probability classes is done by the decision maker; the criteria are not only depending on the type of problem but also on the range of possibilities acceptable to the decision maker, thus related to her perception of risk. Probability and probability classes are assigned to each individual component of the system when dealing with a component-based analysis or to an area/ sector when the analysis is focused on an area with specific and known risk features.

Independently of the type of failures that may take place, they can result in a range of potential consequences not only to the water infrastructure and services but also to other infrastructures. Moreover, consequences can also include socio-economic disruptions and environmental impacts. Therefore, when assessing the risk associated with a specific event, several consequence dimensions should be taken into consideration (Table 1).

Dimension	Type of variables to express relative value in each class
Health and safety	number and severity of injuries number and severity of people affected by disease number of people affected permanently (mortality and disability)
Financial	monetary value; should be a function of the size of utility e.g. annual operating budget (AOB)
Service continuity	Duration of service interruption (availability and compliance with minimum standards); differentiation of type of client affected can be used (residential, hospital, firefighting)
Environmental impacts	Severity e.g. expressed as expected time for recovery (long-term ""/> y years"; mid-term "x to y years"; short-term "w to v months"; rapid recovery "less than w months") Extent (e.g. dimension of area, water quality index, volume or duration of event) Vulnerability (e.g. protected areas, abstraction areas of influence for water supply)
Functional impact on the system	Various performance measures (e.g. population/clients not supplied for a T "/ $>D_{interruption}$; client.hours without supply); thresholds can be associated with legal requirements
Reputation and image	number of complaints; number of times the name of the utility appears in the media, …
Business continuity	damage to materials, service capacity, available human resources to maintain system function and recovery time (e.g. % capacity affected.hours)
Project development	effect on deviation of objectives (e.g. scope, schedule, budget)

Table 1. Dimensions of consequence (adapted from Almeida *et al.*, 2011)

Although other classes of consequences may be adopted, a typical classification might look like this: 1 – insignificant; 2 – low; 3 – moderate; 4 – high; 5 – severe.

The way in which probability and consequence are combined reflects the degree of cautiousness of the analyst, which may vary. Fig. 3 shows a moderate risk perception matrix. A risk matrix should have at least three risk levels (low, medium and high risks) that are to be associated with the acceptance levels of risk: Low or acceptablerisk (green); Medium or tolerable risk (yellow); and High or unacceptablerisk(red)(Almeida *et al.*, 2010).

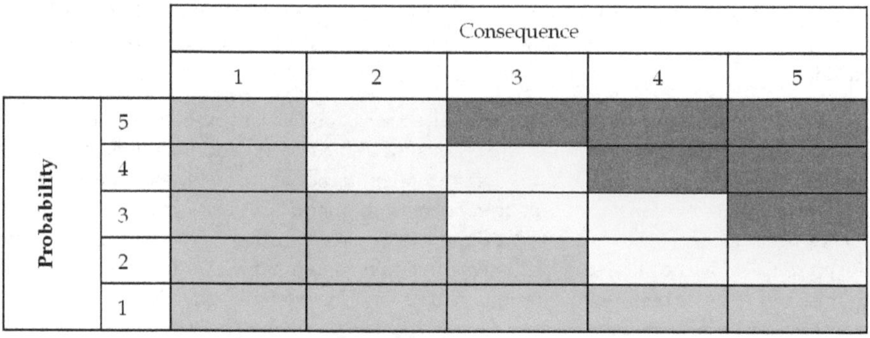

		Consequence				
		1	2	3	4	5
Probability	5					
	4					
	3					
	2					
	1					

Figure 3. Risk matrixadopting a moderate risk perception

5.3. Cost assessment

Cost assessment is the other fundamental axis of analysis for comparing and selecting intervention alternatives in an IAM framework. All relevant costs and revenues items that take place during the analysis horizon and which differ from the *status quo*, should be accounted for, for any of the intervention alternatives considered.

The inclusion in the analysis of cost items that are common in nature and value to all alternatives is optional, as they will not have an effect on the comparison but may be useful in informing it. However, if quantifying the actual net present value or internal rate of return of a financial project is important to the exercise, then all the relevant costs and revenues must be included. In practice, it is often the case that rehabilitation interventions do not affect revenues, and mainly have an effect on system performance, onsystem risk(by affecting system reliability) and on capital and operational costs (e.g., repair costs, complaint management, regulatory or contractual service compliance failure).

In general and simplified terms, the main cost items include:

• Investment costs, expressed as a given amount at a given point in time, and with a given depreciation period (if not linear, a depreciation functionmust be known as well).

- Operational costs, normally organized in three classes: (i) Cost of goods sold; (ii) Supplies & external services; (iii) personnel; operational costs are expressed as annual values, over the analysis period.

- Revenues, either through lump sums occurring at specific points in time (e.g. public subsidies), or distributed over the analysis period (e.g. revenues from tariffs). Revenues are also expressed by their annual value over the analysis period.

Whenever relevant, the costs of planning and designing new assets, as well as disposal costs of assets that reach the end of their service lives, should be included.

Since the end of the analysis horizon does not coincide in general with the end of the service life of most assets, the residual value of all assets at the end of the analysis period must be considered.

Cost-benefit analysis may include not only direct costs and revenues, as described above, but also indirect (i.e., those that are direct costs for a third party) and intangible costs and benefits. However, practice shows that utilities often do not feel comfortable in expressing certain such costs in monetary terms (e.g., increasing public health risk because the water quality does not meet the targets). An option that is recommended by some approaches (Alegre*et al.*, 2011) and successfully implemented in a good number of utilities is to express indirect and costs as performance or risk metrics, and include only direct costs in the cost axis of the analysis.

6. Strategic, tactical and operational planning

Strategic planning needs to be groundedon the utility's vision and mission. It should be built for the entire organization, and it aims at establishing the global and long-term corporate directions.

The first stage is the definition by top management of clear objectives, assessment criteria, metrics to assess them, and finally, targets for every metric. Realistic objectives and targets require proficient knowledge of the context. In general, this is provided by the monitoring and feedback procedures in place. If a utility is preparing a strategic plan for the first time, setting up objectives requires taking into account the available context information, even if not structured and accurate.

The second stage isdiagnosis, consistingof ananalysis of external context (global and stakeholder-specific) and of the internal context (both organizational and infrastructure), anchored in the objectives and targets established. The context evaluation should be carried out through to theplanning horizon. SWOT (strengths-weaknesses-opportunities-threats) analysis is a suitable way to express the results of this stage.

The third stage is the formulation, comparison and selection of strategies that lead to meeting the targets, given the diagnosis. The results should be expressed in a document, the strategic plan, a document that should be synthetic, clear, and effectively disseminated to all relevant internal and external stakeholders.

The implementation of the strategic plan is ensured by a suitable chain of management, where the tactical and operational planners and decision makers play key roles. Implementation should be monitored periodically (in general, annually). Strategic plans should be kept up to date, so that global and long-term directions are known and clear to the entire organization at all times. This may require reviewingand updating every 1/3 to 1/5 of the plan's horizon.

Tactical planning and decision-making should be founded on the strategies and on the strategic objectives and targets. The aim of tactical planning is to define what are the intervention alternatives to implement in the medium term (typically 3 to 5 years). IAM tactical planning is not restricted to infrastructural solutions, as it should also consider the interventions related to operations and maintenance and to other non-infrastructural solutions. Managing the infrastructure has close interdependencies with the management of other assets: human resources, information assets, financial assets, intangible assets. The IAM plan needs to address the non-infrastructural solutions that are critical for meeting the targets and are related to these other types on assets, e.g., investing in a better work orders data system.

The key stages of tactical planning are similar to those described for strategic planning. The objectives, metrics and targets need to be coherent and aligned with the strategic level. Metrics should address all three dimensions of performance, risk and cost.

The diagnosis should be carried out based on the metrics selected, for the present situation and for the planning horizon. Due to the system behavior of the water infrastructures, there is the need to adopt a progressive system-basedscreening progress,aimed at identifying the most problematic areas. In general, the water systems under analysis should be divided into sub-systems, and the metrics assessed for each of them. The most problematic are captured and analyzed in more detail. For those that do not display significant overall problems, there is the need to confirm that they do not have relevant localized problems. If they do, these localized areas need to be retained as well for detailed analysis. This screening process leads to the identification of priority areas of intervention. For these, the diagnosis needs to be more detailed in order for the causes of the problems to be properly understood. The screening process may not apply tonon-infrastructural interventions affecting theentire organization (e.g. organizational changes, IT and information system upgrades).

The next stage is actually producing theplan, and is one of the most work-intensive as it encompasses the demanding engineering processes involved in identifying and developing feasible intervention alternatives for each of the subsystems, and the assessment of their responses over the analysis horizon for the metrics selected. For each subsystem, the intervention alternatives need to be compared, and that alternative which best balances the set of metrics for the chosen objectives, over the long-term, will be selected. The set of best interventions alternatives, compatible with the financial resources that can be mobilized and with the planning horizon, will be included in the tactical plan. The plan must make allowance for the resources needed to implement it.

The detailed diagnosis and the design and analysis of infrastructural and operational intervention alternatives are not trivial tasks and often require the use of sophisticated modeling tools. This is where the more advanced research efforts have been centered, such as mentionedin section 2 (e.g. Skipworth*et al.*, 2002; Sægrov ed., 2005; Sægrov ed., 2006; Malm*et al.*, 2012; Renaud *et al.*, 2011; Alegre and Almeida ed., 2009).

The last stages of tactical planning are the implementation, monitoring and periodic review of the plan. Implementation is materialized via operational management. Monitoring and reviewing are critical for the continuous improvement process. It is recommended that the tactical plan defines their modes, responsibilities and periodicity. Operational IAM planning aims at implementing the interventions selected in the tactical level.

7. Long-term balanced design - carrying urban water systems into the future

As explained before, the performance of individual components is only relevant inasmuch as it contributes to system performance. Some components will have more impact on the system than others, and the behavior of such systems is usually quite complex, giving rise in the last decades to a whole field of expertise devoted to developing and using network analysis models, among the most advanced and useful tools in engineering.

From the viewpoint of infrastructure asset management, the notions of "system design", "preventive maintenance" and "system rehabilitation" should be seen fundamentally as part of the same long-term balanced design process.

Even in those parts of the world where service coverage has reached its effective limit, and designing new systems or system extensions appears to be a thing of the past, it must be realized that design skills and experience are just as needed in carrying present-day systems into the future as they once were in creating the first outlines.

Essentially, investing in a system over a period of time should maximize the performance-risk-cost balance while transforming the system into its ideal for the next 20 or 30 years: that which best serves the strategic objectives defined for the infrastructure as a whole, as explained previously.

If at a strategic IAM level it is common to try to balance conflicting objectives (e.g., improving the environmental sustainability and reducing costs to ensure economic sustainability), at the tactical and operational level, which must be aligned with the former, that is also the nature of the problem: e.g., water supply reliability is commonly achieved through pipeline redundancy, which often causes reduced flow velocities and potentiates water quality issues.

On the other hand, analyzing over long periods of time must account for what is usually a changing context: societal values and expectations evolve; regulations become more demanding; technologies improve; urban areas progress; the climate and the environment fluctuate and change; natural resources become scarcer.

The current emphasis on water-energy efficiency is driven by most of the above factors of change. However, old paradigms are broadly accepted without being questioned. For instance, drinking water networks are still designed in most developed countries to respond to fire flows. Is this the most rational approach? In the Netherlands, for instance, this paradigm is changing. Smaller diameter networks are not only less expensive but also generally behave better in terms of water quality. Firefighting is ensured from a basic trunk main grid. If paradigm shifts occur, rehabilitation interventions need to take them into account.

The fact that most water systems are far from ideal today is a consequence of a growth process that has been forced to react to that changing context over the decades. Most mature systems today are not exactly what they would be if we were to start with a clean slate. Yet, it is common to see preventive maintenance or rehabilitation strategies centered on replacing the pipes with a higher risk of failure with new pipes of the same size. Would it not make sense to try to project the best possible system for a given time horizon – 20, 30 years – and use those very same opportunities of intervention to make the present day system gradually morph into that better design?

The fact is that there are many cases when the waternetworks are adequately and efficiently designed and operated, meeting the hydraulic, water quality and energy targets for the present and for the expected future demands. In these cases, the key driver for rehabilitation is indeed the risk of pipe failure, usually assessed through the combination of failure probability and component importance (in terms of the consequence its failure). Much of the leading-edge theory and practice is tailored for these situations, where the like-for-like replacement strategy fits well.

In classical terms, infrastructures used to be seen as living through a sequence of stages, from the initial design, through constructing new (or extending), operating, maintaining and rehabilitating or replacing by new again. This is indeed the typical AM approach for other types of physical assets. In mature infrastructures, however, all these stages co-exist, and designing new, extending, maintaining or rehabilitating are fundamentally parts of the same process.

The IAM framework introduced in sections 3 to 6induces essentially one approach to the problem, illustrated in Fig. 4 in very simple terms. IAM planning starts from an existing infrastructure and aims at optimizing its behavior over the analysis period, enabling a progressive improvement of the infrastructure condition and functional response. In well-maintained mature infrastructures, this requires that the fair value at the end of the planning horizon is not lower than the initial value.

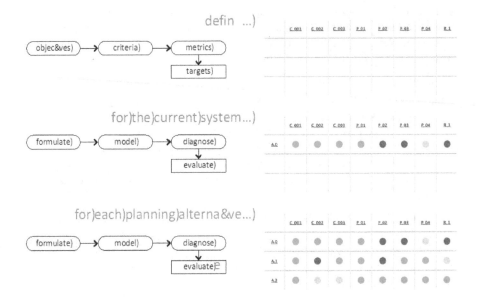

Figure 4. The long-term balanced design planning process

The drawing board on the right-hand side is initially marked out by the green vertical lines, representing the metrics for the criteria chosen to drive the analysis. A thorough diagnosis and assessment of the current system according to those metrics is carried out (represented by the first blue horizontal at the top).

The planning board is then successively populated with the best available planning alternatives (represented by the subsequent blue lines). The intersections represent the assessment of each planning alternative for each metric. The purpose of the process is to fill out the table to the extent possible.

8. Examples from the industry

8.1. Strategic planning in a midsize utility

The vast majority of water utilities in the world serve populations of less than 100 000. Most midsize utilities have room for significant improvement in terms of infrastructure asset management. This specific example arises from Portugal, where the water services regulator enforces a national system for quality of service assessment, and concerns a midsize utility in a developed urban area (more detail can be found in Marques *et al.*, 2012). Service coverage is no longer an issue, but the assets are aging, and the service is not as financially and environmentally efficient as desirable. Quality of service, transparency in investment prioritization and environmental sustainability are the key IAM drivers for the managers.

The utility adopted the objectives and assessment criteria of the regulatory system, as they were deemed adequate for their own internal strategic purposes. Operating exclusively as a retail services utility, they selected the applicable metrics and targets from the regulatory system (Table 2). Each metric is clearly defined, with units, definition, assessment rule and specification of the input variables.

Taking these objectives into account, a SWOT analysis was carried out (Table 3).

Objectives and criteria	Metrics
1. Adequacy of the service provided	
1.1 Service accessibility	Physical accessibility of the service (WS, WW)
	*Economical accessibility of the service (WS, WW)
1.2. Quality of service provided to users	*Service interruptions (WS)
	*Quality of supplied water (WS)
	*Reply to written suggestions and complaints (WS, WW)
	*Flooding occurrences (WW)
2. Sustainability of the service provision	
2.1. Economic sustainability	*Cost coverage ratio (WS, WW)
	Connection to the system (WS, WW)
	*Non-revenue water (WS)
2.2. Infrastructural sustainability	*Adequacy of treatment capacity (WS)
	*Mains rehabilitation (WS)
	*Mains failures (WS)
	*Sewerage rehabilitation (WW)
	*Sewer collapses (WW)
2.3. Physical productivity of human resources	*Adequacy of human resources (WS, WW)
3. Environmental sustainability	
3.1. Efficiency of use of environmental resources	*Energy efficiency of pumping installations (WS, WW)
3.2. Efficiency in pollution prevention	Sludge disposal from the treatment plants (WS, WW)
	*Adequate collected wastewater disposal (WW)
	* Emergency overflow discharges control (WW)
	Wastewater quality tests carried out (WW)
	Compliance with discharge parameters (WW)

WS: water supply services; WW: wastewater services; *adopted by the utility to assess the strategic objectives.

Table 2. Objectives, assessment criteria and metrics of the Portuguese regulatory system

STRENGTHS	WEAKNESSES
- Good information systems on the water supply infrastructures	- Insufficient information systems on wastewater infrastructures
- Sufficient information to assess the water supply systems condition and performance	- Financial restrictions
- Strong competence of human resources	- Inadequate tariffs
- Relation between information systems and work orders	- Poor structural infrastructure condition
	- Poor functional infrastructure performance
	- Insufficient historical records
	- Inadequate quality of data

OPPORTUNITIES	THREATS
- Equipment and technologies available to support IAM	- Portuguese legislation and regulation by ERSAR*
- Portuguese regulation by ERSAR *	(increase in costs)
- Portuguese legislation related with IAM	- Political uncertainties
- Incentives for sustainable use of energy	- Economic crisis and financial restrictions
	- Demographic development uncertainties
	- Illegal cross connections in wastewater systems

* ERSAR: the water and waste services regulator in Portugal

Table 3. SWOT analysis summary

The SWOT analysis results led to the establishment of strategies. For drinking water, the key selected strategies were *Controlwater losses* and *Promote proactive rehabilitation practices*, whereas for wastewater the strategies established were *Reduce untreated wastewater discharges* and *Reduce cross connections and infiltration/inflow in wastewater systems*. The common strategies of both types of services were *Improve infrastructure information systems* and *Increase system reliability*.

8.2. Tactical planning in a midsize utility

Let us put ourselves now in the position of a middle manager of the same utility, in charge of infrastructure planning and rehabilitation for the water supply system. Let us take as an example the strategic objective *Improve theefficiency of use of environmental resources (water and energy)*, as listed in (see criterion 3.1). The utility's networks display undesirable failure rates (pipe breaks) and the energy bill for pumping is higher than would appear reasonable; the network has unflattering water losses and localized pressure problems during peak consumption hours remain.

- How would we act?

- How would we prove that our decisions are effectively addressing the strategic objective?

- How would we quantify the impact of our decisions and of subsequent actions?

In traditional AM practice, we would probably start by gathering an updated and reliable inventory of the existing assets and by compiling as many reliable records as possible of their condition and failure history. We would try to identify the locations where there are pressure problems, and we would also look at pump efficiency and energy consumption. We would probably try to assess the relative importance of each asset. Combining these types of information, we would prioritize interventions within our budget constraints.

This would contribute to answering the first question. What could be done about the other two? Fixing pumps and replacing some pipes will undoubtedly contribute to saving water and energy. But would that maximize the utility of the investment made? A discerning board might be less than satisfied; and the third question would still remain unanswered. They might ask some additional questions:

- Have we satisfactorily dealt with the hydraulic problems? Were we able to allocate levels of service to each individual asset when dealing with pressures and water losses?

- How did we select the sizes and materials of the new pipes?

- Did we assume that the existing network's configurations (e.g., layout and diameters of networks, location and characteristics of storage tanks and pumping stations) are adequate from the energy point of view?

These are the types of issues that a good IAM approach should aim to tackle in a structured, aligned and transparent way.As a basis for tactical planning, this utility took the strategic directions previously defined: objectives, targets and strategies. The following tactical IAM objectives were set:

- Increase system reliability in normal and contingency conditions (see criterion 1.2, Table 2);

- Ensure economic sustainability (see criterion 2.1,Table 2);

- Ensure the infrastructural sustainability of the system (see criterion 2.2,Table 2);

- Decrease water losses (see criterion 3.1,Table 2).

At a first stage of tactical planning, the network was evaluated coarsely in its main subdivisions: trunk main system and supply subsystems (DMAs, or District Metering Areas). The prioritisation of DMAs with higher intervention needs was based on the assessment of the selected metrics for all DMAs, not only for the current situation, but also by assessing the response of the existing systems to the predicted evolution of external factors (e.g., demands, regulation, funding opportunities, economics).

DMA 542 was in this high priority group, since it failed to comply with most tactical targets. It supplies a stable and heterogeneous urban area, comprising new and old residential buildings, schools, shops and some commercial areas. It supplies approximately 10,000 people (4,388 contracts) with a network of approximately 12.5 km of total pipe length, 40% of which in asbestos cement and the remainder in more recent plastic materials. Water is supplied by gravity from a service tank at elevation 185 m, and the lowest ground elevationis 107 m.

The tactical plan was designed for a 5-year planning horizon (2011-2016). Any envisaged interventions will have to be scheduled over this period. However, the evaluation was carried out over a 20-year analysis horizon in order to ensure that the interventions planned are the best compromise both in the medium-and in the long-term (Alegre*et al.*, 2011). The available investment budget for this DMA allows for the replacement of approximately 1 km of pipeline per year, for 5 years. Reference assessment timesteps were considered at years 0, 1, 2, 3, 4, 5, 10, 15 and 20 (i.e., 2011 to 2031).

Since this example involves only alternatives related to physical intervention in the infrastructure, compliance with the above-mentioned tactical IAMobjectives was assessed through the following performance, risk and cost metrics:

- **Inv**: *investment cost*, measured through the net present value at year 0 of the investments made during the 5-year plan.

- **IVI**: *infrastructure value index (IVI*, the ratio between the current value and the replacement value of the infrastructure (Alegre and Covas, 2010); it should ideally be close to 0.5.

- P_{min}: *minimum pressure under normal operation index*, measuring compliance with the minimum pressure requirementsat the demand locations.

- P_{min}*: *minimum pressure under contingency conditions index,*measuring compliance with the minimum pressure requirementsat the demand locations when the normal supply source point to this DMA fails and an alternative entry point is activated.

- **AC**: *percentage of total pipe length in asbestos cement*; although this metric may seem unconventional as a performance indicator, it was selected as a proxy for system resilience, reliability and ease of maintenance (or the lack thereof), given the poor track record of the aging asbestos cement pipes in this utility.

- **RL**: *real losses per connection*, as defined in the IWA performance indicator system (Alegre*et al.*, 2006).

- **UnmetQ**: *risk of service interruption*. This reduced service metric is given by the expected value of unmet demand over 1-year period. The risk of service interruption associated to a specific pipe depends on the likelihood of its failure and on its consequence on the actual service. This risk is calculated for each pipe as a combination of failure probabilityand component importance.

The values of the metrics were further divided into 3 ranges (good, fair and poor) according to the thresholds set by the utility, based on the experience of their key staff (Table 4).

The diagnosis of the situation at year 0 using the assessment metrics and associated reference values pointed to the following problems:

- *Reliability of the system:* insufficient pressure in normal conditions at some locations; high pipe failure rates; low system resilience in contingency operation conditions.

- *Infrastructural sustainability:* poor condition (high failure rates) of asbestos cement pipes.

- *Water losses:* high leakage levels.

	Good (green)	Fair (yellow)	Poor (red)
Inv (cost units)	[0, 350[	[350, 450[	[450, ∞[
IVI (-)	]0.45, 0.55[	[0.30, 0.45[; [0.55, 0.70[	[0, 0.30]; [0.70, 1]
Pmin (-)	[3, 2[	[2, 1[	[1, 0]
Pmin* (-)	[3, 2[	[2, 1[	[1, 0]
AC (%)	[0, 9[	[9, 15[	[15, 100]
RL (l connection⁻¹ day⁻¹)	[0, 100[	[100, 150[	[150, ∞[
UnmetQ (m³/year)	[0, 20[	[20, 30[	[30, 100]

Table 4. Multi-criteria reference values

Several system-driven solutions and like-for-like replacement solutions, within the available budget, were analysed (Marques *et al.*, 2011) and designed to solve or mitigate the problems identified in the diagnosis, both in-house and through external consultants. The final set of alternative solutions were summarized as follows (including retaining the *status quo*):

- **Alternative A0** (*status quo*, or base case): corresponds to keeping the existing network as it is, and retaining the current reactive capital maintenance policy (which in the present case was based on repairs after break only).

- **Alternative A1** (*like-for-like replacement*): an IAM project consisting of a prioritized list of pipes to be replaced by the same-diameter HDPE pipes. The prioritized list was developed externally to the AWARE-P software, following a like-for-like replacement strategy, using pipe failure and consequence analysis (as in FAIL/CIMP) and an ELECTRE TRI decisional method, and taking into consideration 3rd-party coordination.

- **Alternative A2** (*system-driven solution*): an IAM project based on an *ideal* redesign for the network, as if it were built from scratch for the present-day context – significantly different from the actual current network, which was designed and constructed from the 1940s onwards. This ideal redesign, heavily backed by network modelling, driven by performance and risk assessments, is viewed by the utility as a future target reference, to be gradually reached by incrementally changing individual pipes as they are replaced, and by making some key layout modifications. It addresses the same pipes targeted in A1, but replaces them with new pipes of optimal diameter (often smaller, as the original network has overcapacity in places); in Year 5, a new 625 m-long pipeline connecting to a neighbouring DMA is introduced in order to improve reliability of supply in emergency situations.

The assessment of the three alternatives was carried out for the 5-year planning horizon and for a 20-year analysis horizon.Table 5illustrates the results of the selected metrics for the three alternatives at Year 5. Fig. 5shows snapshots of the 3D view of results, with time, assessment metrics, and alternatives depicted respectively along the left, right and vertical axes. Themajority of the assessment metrics are constant after year 5 (with the exception of IVI and UnmetQ), due to the adoption of a constant demandscenario (this is a very stable residential area),and to having assumed negligible growth of O&M costs. In this case, the comparison and selection of alternatives can be basedon the assessmentfor Year 5.

Alternatives	Assessment metrics						
	Inv (c.u.)	IVI (-)	Pmin(-)	Pmin * (-)	AC (%)	RL (l conn⁻¹ day⁻¹)	UnmetQ (m³/year)
A0	0	0.47	2.88	0.00	37.2	116	36
A1	274	0.73	2.88	0.00	1.5	52	22
A4	350	0.70	2.99	2.99	8.5	54	18

Table 5. Case study: results obtained from the evaluation of three alternatives at year 5

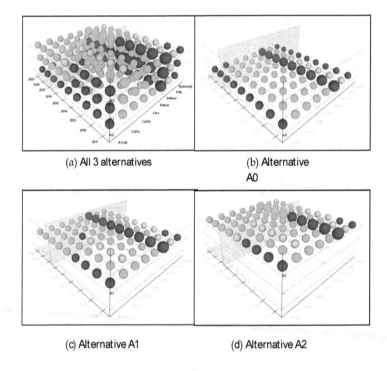

(a) All 3 alternatives   (b) Alternative A0

(c) Alternative A1   (d) Alternative A2

Figure 5. Metric results expressed as a 3D *cube*; left axis: time; right axis: metrics; vertical axis: alternatives.

Experience shows that it is often less costly simply to repair pipes and pay for the water lost in leakage than to invest in the rehabilitation of the system. This was confirmed here by looking at alternative A0 at year 5. However, for the remainder of the analysis period (yrs. 6-20) the problems identified in the diagnosis become increasingly evident, throughpoorer network reliability andmoderate water losses that tend to intensify due to normal wear.

The results for A1 show that it is generally better than A0 is terms of infrastructural sustain-ability, water losses and risk (IVI, AC and UnmetQ). Investment is of course higher than in A0, but within the available budget. However, A1 perpetuates the design deficiencies inher-ent tothe existing system (A0).

Alternative A2 aims at realistically and progressively bring the existing network to a config-uration closer to the ideal. Its resilience is improved when compared to A0 and A1, as it re-inforcesthe options for supplying the network from an alternative supply point. Investment costs are higher than for A1 (350 vs. 274cost units). The percentage of asbestos cement pipes is also significantly reduced (to 8.5%, from 37% for A0). This alternative displays the best all-round long-term balance of performance, risk and cost, as expressed by metrics that reflect the tactical objectives, in full alignment with the utility's strategic objectives.

8.3. Benefits of using a structured IAM approach in the example utility

The adoption of a structured IAM approach inthe utility illustrated by this example provid-ed proficient answers to all the questions initially formulated:

- Using a coherent and aligned system of objectives, metrics and metrics enables the IAM manager to show that the decisions are effectively addressing the strategic objectives, and to quantify their impact.

- The hydraulic problems were duly taken into account by splitting the whole system into subsystems and analysing in more detail, including in hydraulic terms, the most problem-atic ones.

- The selection of sizes and materials forthe new pipes was driven by the ability of the existing network in meeting current and future needs and in minimizing energy consumption.

9. Concluding remarks

Infrastructure asset management of urban water infrastructures will be increasingly critical in the coming decades. In industrialized countries, particularly those affected by World War II, the heavy investments in new systems carried out in the 1950's, 1960's and 1970's are ag-ing fast, partly due to inadequate or deferred capital maintenance. This places an additional demand for efficiency in planning for the future.In developing regions, the shortageof finan-cial and technical resources further add to the need for their well-judged, efficient usein a long-term perspective.

With the current lack of planning and capital maintenance, the services that are taken for granted in many societies are placed into an increased risk of failure, at least from the view-point of the levels of service currently provided.

Regardless of their size, complexity and level of maturity or development, water utilities need to implement structured IAM approaches that may ensure the sustainable management of their systems.Thereare some key recommendations to be taken into account when implementing an IAM program:

- IAM is all about people – successful implementation requires:

leadership;

co-ordination;

collaboration;

corporate culture acceptance;

team building;

motivation;

commitment;

know-how.

- IAM is not implemented overnight. It is an incremental, step by step process that must be keptas simple as possible with a long-term view. The structured approach recommended in thischapter aims at identifying intervention priorities, includingnew organizational procedures, data management and decision making processes.

- Reliable data are the foundation of successful IAM. Before investing on new data collection, it is vital to get the most out of the existing data, through proficient recycling, quality control, analysis and interpretation.

- As superfluous as the statement may seem, IAM is not solved or even set in motion by acquiring a software application.

- IAM is an internal process in a utility. Although external expert advice is valuable, it should be seen as a contribution to an internally driven effort, e.g., in capacity building or to sort out specific technical issues problems, such as when engineering consultancy is brought in to develop and advise on infrastructural alternatives to solve given issues. IAM should not be outsourced.

- Water utilities have many common problems and difficulties. Sharingthem, and any solutions, among peers has always proved to be enriching, effective and highly motivational.

- Make it happen – start today!

Author details

Helena Alegre* and Sérgio T. Coelho*

LNEC - National Civil Engineering Laboratory, Lisbon,, Portugal

References

[1] Alegre, H., Almeida, M. C., & ed, . (2009). Strategic asset management of water and wastewater infrastructures. IWA Publishing,, 97843391869536.

[2] Alegre, H., Almeida, M. C., Covas, D. I. C., Cardoso, M. A., & Coelho, S. T. (2011). Integrated approach for infrastructure asset management of urban water systems". International Water Association 4th Leading Edge Conference on Strategic Asset Management, 2730de Setembro, Mülheiman der Ruhr, Germany, 10.

[3] Alegre, H., & Covas, D. (2010). Infrastructure asset management of water services. (in Portuguese). Technical Guide n.16. ERSAR, LNEC, IST, Lisboa,, 472, 978-9-89836-004-5.

[4] Alegre, H., Covas, D. I. C., Coelho, S. T., Almeida, M. C., Cardoso, M. A., & (2012, . An integrated approach for infrastructure asset management of urban water systems. Water Asset Management International 8., 2(2012), 10-14.

[5] Alegre, H., Baptista, J. M., Cabrera, J. R., , E., Cubillo, F., Duarte, P., Hirner, W., Merkel, W., & Parena, R. (2006). Performance indicators for water supply services,. second edition, Manual of Best Practice Series, IWA Publishing, London,, 184-3-39051-530-5.

[6] Alegre, H., Hirner, W., Baptista, J. M., & Parena, R. (2000). Performance indicators for water supply services,. 1st edition, Manual of Best Practice Series, IWA Publishing, London,, 190-0-22227-216-0.

[7] Almeida, M. C., & Cardoso, M. A. (2010). Infrastructure asset management of wastewater and stormwater services. (in Portuguese). Technical Guide n.17. ERSAR, LNEC, IST, Lisboa, 978-9-89836-005-2.

[8] Almeida, M. C., Leitão, J. P., & Borba, R. (2010). AWARE-P Development reports: Risk Assessment Module. Internal report of the AWARE-P project (classified document, in Portuguese).

[9] Almeida, M. C., Leitão, J. P., & Coelho, S. T. (2011). Risk management in urban water infrastructures: application to water and wastewater systems. In Almeida, B., Gestão da Água, Incertezas e Riscos: Conceptualização operacional (Water management, uncertainty and risks: operational conceptualisation). Esfera do Caos, Lisbon, Portugal (in Portuguese).

[10] ASCE. (2009). Report Card for America's Infrastructure Advisory Council, American Society of Civil Engineers,, 978-0-78441-037-0.

[11] Brown, R. E., & Humphrey, B. G. (2005). Asset management for transmission and distribution. *Power and Energy Magazine*, IEEE,, 3(3), 39.

[12] Burns, P., Hope, D., & Roorda, J. (1999). Managing infrastructure for the next generation. Automation in Construction,, 8(6), 689.

[13] Cabrera, E., & Pardo, M. A. eds.)((2008). Performance Assessment of Urban Infra-structure Services: drinking water, wastewater and solid waste,. IWA Publishing, 978-1-84339-191-3

[14] Cardoso, M. A., Silva, , , M. S., Coelho, S. T., Almeida, M. C., & Covas, D. (2012). Urban water infrastructure asset management- a structured approach in four water utilities,. *Water Science & Technology*, (2012), IWA Publishing (in press).

[15] Clark, R. M., Carson, J., Thurnau, R. C., Krishnan, R., , R., & Panguluri, S. (2010). Condition assessment modeling for distribution systems using shared frailty analy-sis. *Journal AWWA.*, American Water Works Association, Denver, CO,, 102(7), 81-91.

[16] Coelho, S. T., & Vitorino, D. (2011). AWARE-P: a collaborative, system-based IAM planning software. IWA 4th Leading Edge Conference on Strategic Asset Manage-ment,, 27-30, September, MülheimAn Der Ruhr, Germany.

[17] Hughes, D. M. (2002). Assessing the future: Water utility infrastructure management. *AWWA,* , USA, 644.

[18] IAM/BSI (2008). PAS Asset management, Part 1: Specification for the optimized man-agement of physical infrastructure assets (PAS 55-1); Part 2: Guidelines for the appli-cation of PAS 55-1 (PAS 55-2),. Institute of Asset Management & British Standards Institution- Business Information. , 55.

[19] INGENIUM, IPWEA. (2011). International infrastructure management manual, ver-sion 4.0. Association of Local Government Engineering NZ Inc (INGENIUM) and the Institute of Public Works Engineering of Australia (IPWEA),, 277-0-00007-232-8.

[20] ISO. (2012a). Asset management- Overview, principles and terminology,. ISO/CD 55000.2, ISO/TC 251/WG 1.

[21] ISO. (2012b). Asset management- Management systems- Requirements,. ISO/CD 55001.2, ISO/TC 251/WG 2.

[22] ISO. (2012c). Asset management- Management systems- Guidelines for the applica-tion of ISO 55001,. ISO/CD 55002.2, ISO/TC 251/WG 2.

[23] ISO 24510:. (2007). Activities relating to drinking water and wastewater services-Guidelines for the assessment and for the improvement of the service to users.

[24] ISO 24511: (2007). Activities relating to drinking water and wastewater services-Guidelines for the management of wastewater utilities and for the assessment of drinking water services.

[25] ISO 24512: (2007). Service activities relating to drinking water and wastewater-Guidelines for the management of drinking water utilities and for the assessment of drinking water services.". Intern. Org. for Standardization, Geneva.

[26] Malm, A., Ljunggren, O., Bergstedt, O., Pettersson, T. J. R., & Morrison, G. M. (2012). Replacement predictions for drinking water networks through historical da-ta,. *Water Research*, 4(6), 2012-2149.

[27] Marques, M. J., Saramago, A. P., Silva, M. H., Paiva, C., Coelho, S., Pina, A., Oliveira, S. C., Teixeira, J. P., Camacho, P. C., Leitão, J. P., & Coelho, S. T. (2012). Rehabilitation in Oeiras&Amadora: a practical approach,. *Water Asset Management International*, (in press).

[28] Matos, R.; Cardoso, M.A.; Ashley, R; Duarte, P.; Schulz A (2003). Performance indicators for wastewater services, Manual of Best Practice Series, IWA Publishing, ISBN: 9781900222907 (192 p.).

[29] Renaud, E., Le Gat, Y., & Poulton, M. (2011). Using a break prediction model for drinking water networks asset management: From research to practice,. International Water Association 4th Leading Edge Conference on Strategic Asset Management, 27 - 30September, Mülheiman der Ruhr, Germany.

[30] Sægrov, S., & ed, . (2005). CARE-W- Computer Aided Rehabilitation for Water Networks. EU project: EVK1CT-2000-00053, IWA Publishing,, 184-3-39091-420-8.

[31] Sægrov, S., & ed, . (2006). CARE-S- Computer Aided Rehabilitation for Sewer and StormwaterNetworks. IWA Publishing,, 184-3-39115-514-0.

[32] Sjovold, F.; Conroy, P.; Algaard, E. (2008). Performance assessment of urban infrastructure services: the case of water supply, wastewater and solid waste (146 p.) ISBN: 978-82-536-1010-6

[33] Skipworth, P., Engelhardt, M., Cashman, A., Savic, D., Saul, A., & Walters, G. (2002). Whole life costing for water distribution network management,. Thomas Telford Limited,, 072-7-73166-121-6.

[34] Sneesby, A. (2010). Sustainable infrastructure management program learning environment (SIMPLE). Sustainable Infrastructure and Asset Management Conference. Australian Water Association. 23-24November 2010, Sydney, Australia (CD).

[35] UKWIR (2003). A common framework for capital investment planning,. UK Water Industry Research, ReinoUnido, Regulatory Report 02RG/05/3 (4 Vols.), http://www.ukwir.co.uk/ukwirlibrary/90848,ref. Março 2007.

[36] US EPA. (2012). Condition Assessment Technologies for Water Transmission and Distribution Systems,. Publication No. EPA/600/R-12/017.

[37] Vitorino, D., Coelho, S. T., Alegre, H., Martins, A., Leitão, J. P., & Silva, M. S. (2012). AWARE-P software documentation,. AWARE-P project,, www.baseform.org.

Chapter 2

Guidelines for Transient Analysis in Water Transmission and Distribution Systems

Ivo Pothof and Bryan Karney

Additional information is available at the end of the chapter

1. Introduction

Despite the addition of chlorine and potential flooding damage, drinking water is not generally considered a hazardous commodity nor an overwhelming cost. Therefore, considerable water losses are tolerated by water companies throughout the world. However, more extreme variations in dry and wet periods induced by climate change will demand more sustainable water resource management. Transient phenomena ("transients") in water supply systems (WSS), including transmission and distribution systems, contribute to the occurrence of leaks. Transients are caused by the normal variation in drinking water demand patterns that trigger pump operations and valve manipulations. Other transients are categorised as incidental or emergency operations. These include events like a pumping station power failure or an accidental pipe rupture by external forces. A number of excellent books on fluid transients have been written (Tullis 1989; Streeter and Wylie 1993; Thorley 2004), which focus on the physical phenomena, anti-surge devices and numerical modelling. However, there is still a need for practical guidance on the hydraulic analysis of municipal water systems in order to reduce or counteract the adverse effects of transient pressures. The need for guidelines on pressure transients is not only due to its positive effect on water losses, but also by the contribution to safe, cost-effective and energy-saving operation of water distribution systems. This chapter addresses the gap of practical guidance on the analysis of pressure transients in municipal water systems.

All existing design guidelines for pipeline systems aim for a final design that reliably resists all "reasonably possible" combinations of loads. System strength (or resistance) must sufficiently exceed the effect of system loads. The strength and load evaluation may be based on the more traditional allowable stress approach or on the more novel reliability-based limit state design. Both approaches and all standards lack a methodology to account for dynamic hydraulic loads (i.e., pressure transients) (Pothof 1999; Pothof and McNulty 2001). Most of the current standards simply state that dynamic internal pressures should not exceed the design pressure with a certain factor, duration and occurrence frequency. The Dutch standard NEN 3650 (Requirements for pipeline systems) includes an appendix that provides some guidance on pressure transients (NEN 2012).

One of the earliest serious contributions to this topic was the significant compilation of Pejovic and Boldy (1992). This work not only considered transient issues such as parameter sensitivity and data requirements, but usefully classified a range of loading conditions that accounted for important differences between normal, emergency and catastrophic cases, and the variation in risk and damage that could be tolerated under these different states.

Boulos *et al.* (2005) introduced a flow chart for surge design in WSS. The authors address a number of consequences of hydraulic transients, including maximum pressure, vacuum conditions, cavitation, vibrations and risk of contamination. They proposed three potential solutions in case the transient analysis revealed unacceptable incidental pressures:

1.   Modification of transient event, such as slower valve closure or a flywheel;

2.   Modification of the system, including other pipe material, other pipe routing, etc.; and

3.   Application of anti-surge devices.

Boulos *et al.* list eight devices and summarise their principal operation. They do not provide an overview of the scenarios that should be included in a pressure transient analysis. Jung and Karney (2009) have recognised that an *a priori* defined design load does not necessarily result in the worst-case transient loading. Only in very simple systems can the most critical parameter combination can be defined *a priori* (Table 4). In reality, selecting appropriate boundary conditions and parameters is difficult. Further, the search for the worst case scenario, considering the dynamic behaviour in a WSS, is itself a challenging task due to the complicated nonlinear interactions among system components and variables. Jung and Karney (2009) have extended the flow chart of Boulos *et al.* (2005), taking into account a search for the worst-case scenario (Figure 1). They propose to apply optimisation tools to find the worst-case loading and a feasible set of surge protection devices.

Automatic control systems have become common practice in WSS. Since WSS are spatially distributed, local control systems may continue in normal operating mode, after a power failure has occurred somewhere else in the system. The control systems may have a positive or negative effect on the propagation of hydraulic transients. On the other hand, the distributed nature of WSS and the presence of control systems may be exploited to counteract the negative effects of emergency scenarios. Therefore, existing guidelines on the design of WSS must be updated on a regular basis in order to take these developments into account.

Typical design criteria for drinking water and wastewater pipeline systems are listed in section 2. Section 3 presents a systematic approach to the surge analysis of water systems. This approach focuses on guidelines for practitioners. The key steps in the approach include the following: preconditions for the surge analysis; surge analysis of emergency scenarios without provisions; sizing of anti-surge provisions and design of emergency controls; evaluation of normal operations and design of control systems. The approach has been applied successfully by Both Deltares (formerly Delft Hydraulics) and HydraTek and Associates Inc. in numerous large water transmission schemes worldwide. Especially the integrated design of surge provisions and control systems has many benefits for a safe, cost-effective and energy-efficient operation of the water pipeline system. Section 4 summarises the key points of this paper.

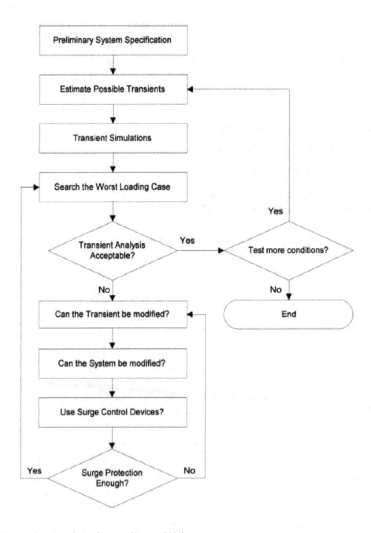

Figure 1. Pressure Transient design (Jung and Karney 2009).

2. Pressure transient evaluation criteria for water pipelines

In any transient evaluation, pressure is the most important evaluation variable, but certainly not the only one. Component-specific criteria must be taken into account as well, such as a minimum fluid level in air vessels, maximum air pressure during air release from an air valve or the maximum fluid deceleration through an undamped check valve.

The maximum and minimum allowable pressure is directly related to the pressure rating of the components. Thin-walled steel and plastic pipes are susceptible to buckling at a combination of external pressure and minimum internal pressure.

The design pressure for continuous operation is normally equal to the pressure rating of the system. During transient events or emergency operation, the system pressure may exceed the design pressure up to a certain factor of the design pressure. Table 1 provides an overview of maximum allowable incidental pressure (MAIP) in different national and international codes and standards.

Code	Maximum Incidental Pressure Factor [-]
DVGW W303:1994 (German guideline)	1.00
ASME B31.4 (1992), IS 328, BS 8010, ISO CD 16708:2000	1.10
NEN 3650-1:2012	1.15
BS 806	1.20
Italian ministerial publication	1.25 – 1.50

Table 1. Overview of maximum allowable incidental pressures (MAIP) in international standards, expressed as a factor of the nominal pressure class.

The minimum allowable pressure is rarely explicitly addressed in existing standards. The commonly accepted minimum incidental pressure in drinking water distribution systems is atmospheric pressure or the maximum groundwater pressure necessary to avoid intrusion at small leaks. If the water is not for direct consumption, negative pressures down to full vacuum may be allowed if the pipe strength is sufficient to withstand this condition, although tolerance to such conditions varies with jurisdiction. Full vacuum and cavitation can be admitted under the condition that the cavity implosion is admissible. Computer codes that are validated for cavity implosion must be used to determine the implosion shock. The maximum allowable shock pressure is 50% of the design pressure. This criterion is based on the following reasoning: The pipeline (including supports) is considered a single-mass-spring system for which a simplified structural dynamics analysis can be carried out. The ratio of the dynamic response (i.e., pipe wall stress) to the static response is called the dynamic load factor (DLF). The dynamic load factor of a mass-spring system is equal to 2. It is therefore recommended that a maximum shock pressure of no more than 50% of the design pressure be allowed. This criterion may be relaxed if a more complete Fluid-Structure-Interaction (FSI) simulation is performed for critical above-ground pipe sections.

3. Systematic approach to pressure transient analysis

The flow chart in Figure 2 integrates the design of anti-surge devices and distributed control systems. It is emphasised that a surge analysis is strongly recommended upon each modification to an existing system. The systematic approach also applies to existing systems.

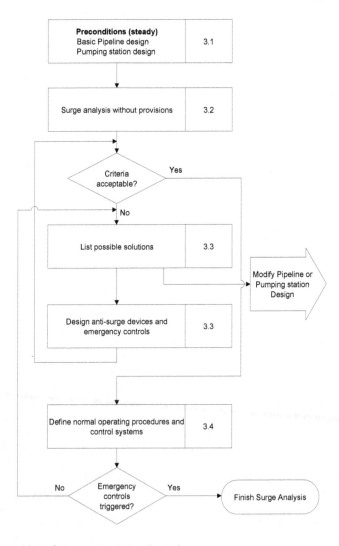

Figure 2. Integrated design for pressure transients and controls.

Because system components are tightly coupled, detailed economic analysis can be a complex undertaking, However, the net present value of anti-surge equipment may rise to 25% of the total costs of a particular system. Therefore, the systematic approach to the pressure transient analysis is preferably included in a life cycle cost optimisation of the water system, because savings on investment costs may lead to operation and maintenance costs that exceed the net present value of the investment savings.

3.1. Necessary information for a pressure transient analysis

The phenomenon of pressure transients, surge or water hammer is defined as the simultaneous occurrence of a pressure and velocity changes in a closed conduit. Water hammer may occur in both long and short pipes. The larger and faster the change of velocity, the larger the pressure changes will be. In this case, 'fast' is not an absolutely term, but can only be used relative to the pipe period, that is, relative to the pipe's internal communications. The most important parameters for the magnitude of transient pressures are:

- Velocity change in time, Δv (m/s) (or possibly the pressure equivalent)

- Acoustic wave speed, c (m/s)

- Pipe period, T (s)

- Joukowsky pressure, Δp (Pa)

- Elevation profile

The acoustic wave speed c is the celerity at which pressure waves travel through pressurised pipes. The wave speed accounts for both fluid compressibility and pipe stiffness: the more elastic the pipe, the lower the wave speed. In fact, all phenomena that create internal storage contribute to a reduction of wave speed. Since air is much more compressible than water, air bubbles reduce the wave speed considerably, but this is the primary positive effect of air in pipelines. The negative consequences of air in water pipelines, particularly in permitting or generating large velocity changes, can greatly exceed this positive effect in mitigating certain transient changes; thus, as an excellent precaution, free or mobile air must generally be avoided in water systems whenever possible and cost-effective. The maximum acoustic wave speed in an excavated water tunnel through rocks is 1430 m/s and drops to approximately 1250 m/s in steel, 1000 m/s in concrete and ductile iron, 600 m/s in GRP, 400 m/s in PVC and about 200 m/s in PE pipes.

$$c = \frac{1}{\sqrt{\rho\left(\dfrac{C_1 D}{eE} + \dfrac{1}{K}\right)}} \tag{1}$$

where:

c = Acoustic wave speed (m/s)

E = Young's modulus of pipe material (N/m²)

K = Bulk modulus of fluid (N/m²)

ϱ = Fluid density (kg/m³)

D = Pipe diameter (m)

e = Wall thickness (m) and

C_1 = Constant depending on the pipe anchorage (order 1).

The acoustic wave speed in water pipelines is shown in Figure 3.

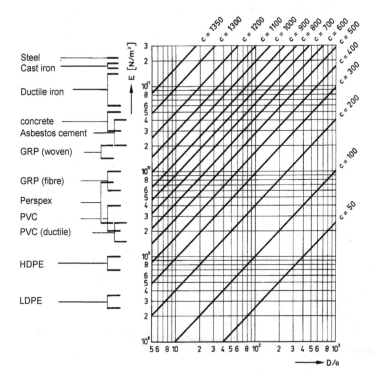

Figure 3. Graph of acoustic wave speed in water pipelines in relation to pipe material (E) and wall thickness (D/e).

The pipe period T [s] is defined as the time required for a pressure wave to travel from its source of origin through the system and back to its source. For a single pipeline with length L:

$$T = 2L/c \qquad\qquad (2)$$

This parameter defines the natural time scale for velocity and pressure adjustments in the system.

Only after the pipe period the pressure wave will start to interact with other pressure waves from the boundary condition, such as a tripping pump or a valve closure. Any velocity change Δv within the pipe period will result in a certain "practical maximum" pressure, the so-called Joukowsky pressure, Δp.

$$\Delta p = \pm \rho \cdot c \cdot \Delta v \tag{3}$$

A slightly more conservative assessment of the maximum transient pressure includes the steady friction head loss $\Delta p_s = \rho g \Delta H_s$.

$$\Delta p = \pm \left(\rho \cdot c \cdot \Delta v + \rho g \Delta H_s \right) \tag{4}$$

All these parameters follow directly from the basic design. The maximum rate of change in velocity is determined by the run-down time of a pump or a valve closure speed. The pump run-down time is influenced by the polar moment of inertia of the pump impeller, the gear box and motor. The full stroke closure time of valves may be increased in order to reduce the rate of velocity change.

Pressure waves reflect on variations of cross-sectional area (T-junctions, diameter changes, etc.) and variation of pipe material. All these parameters must be included in a hydraulic model.

Finally, the elevation profile is an important input, because extreme pressures typically occur at its minimum and maximum positions.

3.2. Emergency scenarios without anti-surge provisions

A pressure transient analysis or surge analysis includes a number of simulations of emergency scenarios, normal operations maintenance procedures. The emergency scenarios may include:

- Complete pump trip

- Single pump trip to determine check valve requirements

- Unintended valve closure; and

- Emergency shut-down procedures.

A pump trip without anti-surge provisions causes a negative pressure wave traveling into the WSS. If the downstream boundary is a tank farm or large distribution network, then the reflected pressure wave is an overpressure wave. If the check valves have closed within the pipe period, then the positive pressure reflects on the closed check valves by doubling the positive pressure wave (Figure 4). In this way, the maximum allowable pressure may be exceeded during a pump trip scenario.

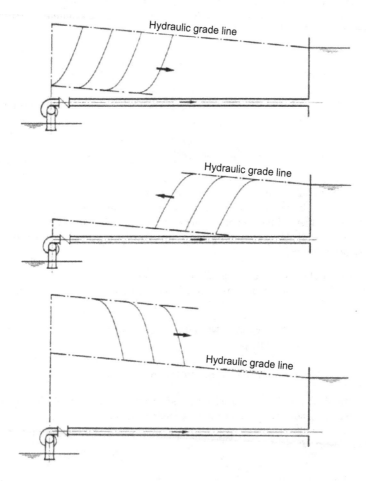

Figure 4. Pressure wave propagation following a pump trip

Check valves will generally close after pump trip. The transient closure of a check valve is driven by the fluid deceleration through the check valve. If the fluid decelerates quickly, an undamped check valve will slam in reverse flow. Fast-closing undamped check valves, like a nozzle- or piston-type check valve, are designed to close at a very small return velocity in order to minimize the shock pressure. Ball check valves are relatively slow, so that their application is limited to situations with small fluid decelerations.

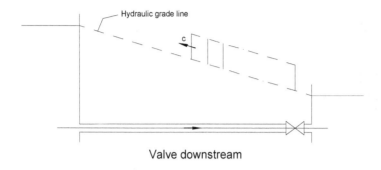

Valve downstream

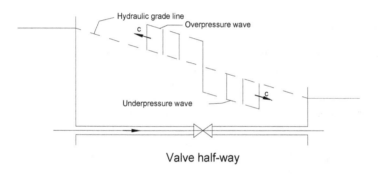

Valve half-way

Figure 5. Pressure wave propagation following valve closure

Emergency closure of a line valve creates a positive pressure wave upstream and negative pressure wave downstream of the valve. Although the total closure time may well exceed the characteristic pipe period, the effective closure may still occur within one pipe period, so that the Joukowsky pressure shock may still occur. The effective closure is typically only 20% of the full stroke closure time, because the valve starts dominating the total head loss when the valve position is less than 20% open (e.g., Figure 6). If a measured capacity curve of the valve is used, simulation software will deliver a reliable evolution of the discharge and transient pressures in the WSS.

Figure 6 shows an example of a butterfly valve at the end of a 10 km supply line (wave speed is 1000 m/s). A linear closure in 5 pipe periods (100 s) shows that the pressure rises only during the last 30% of the valve closure. Therefore the pressure rise is almost equal to the Joukowsky pressure. A two-stage closure, with a valve stroke from 100% to 30% open in 1 pipe period (20 s), shows a more gradual pressure rise during the closing procedure and a lower peak pressure.

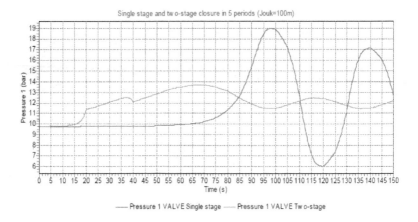

Figure 6. Single and two-stage valve in 5 pipe periods (100 s)

In general, for each scenario multiple simulations must be carried out to determine the extreme pressures and other hydraulic criteria. Scenario variations may include flow distributions, availability of signal transfer (wireless or fiber-optic cable) for the control system and parameter variations. For example, the minimum pressure upon full pump trip will be reached in a single pipeline, if the maximum wall roughness value is used. If an air vessel is used as an anti-surge device, the minimum wall roughness and isothermal expansion must be applied to determine the minimum water level in the air vessel. Adiabatic pocket expansion in air vessels must be applied for other scenarios. The selection of input parameters so that the extreme hydraulic criterion values are computed is called a conservative modeling approach (Pothof and McNulty 2001). The proper combination of input parameters can be determined *a priori* for simple (single pipeline) systems only. Table 4 provides some guidance on the conservative modeling approach.

In more realistic situations a sensitivity analysis is required to determine the worst case loading. A more recent development for complex systems is to combine transient solvers with optimization algorithms to find the worst case loading condition and the appropriate protection against it (Jung and Karney 2009).

In most cases, the emergency scenarios result in inadmissible transient pressures. Possible solutions include modifications to the system or transient event (e.g., slower valve closure), anti-surge devices, emergency controls, or a combination of the above. The solutions will be discussed in more detail in the next section.

3.3. Design of anti-surge devices and emergency controls

In order to mitigate inadmissible transient pressures, hydraulic design engineers have four different management options at their disposal:

1. System modifications (diameter, pipe material, elevation profile, etc.);

2. Moderation of the transient initiation event;

3. Emergency control procedures; and/or

4. Anti-surge devices.

3.3.1. System modifications

Measure 1 is only feasible in an early stage. A preliminary surge analysis may identify cost-effective measures for the surge protection that cannot later be incorporated. If, for example, inadmissible pressures occur at a local high point that seem difficult to mitigate, the pipe routing may be changed to avoid the high point. Alternatively, the pipe may be drilled through a slope to lower the maximum elevation.

Selection of a more flexible pipe material reduces the acoustic wave speed. Larger diameters reduce the velocities and velocity changes, but the residence time increases, which may render this option infeasible due to quality concerns.

A cost-benefit analysis is recommended to evaluate the feasibility of these kinds of options.

3.3.2. Moderating the transient initiation event

A reduction of the rate of velocity change will reduce the transient pressure amplitude. A variable speed drive or soft start/stop functionality may be effective measures for normal operations, but their effect is negligible in case of a power failure. A flywheel increases the polar moment of inertia and thereby slows down the pump trip response. It should be verified that the pump motor is capable of handling the large inertia of the flywheel during pump start scenarios. Experience shows that a flywheel is not a cost-effective option for pumps that need to start and stop frequently.

If inadmissible pressures are caused by valve manipulations, the valve closure time must be increased. The velocity reduction by a closing valve is not only influenced by the valve characteristic, but also by the system. The valve resistance must dominate the total system resistance before the discharge is significantly reduced. Therefore, the effective valve closure time is typically 20% to 30% of the total closure time. A two-stage closure, or the utilization of a smaller valve in parallel, may permit a rapid initial stage and very slow final stage as an effective strategy for an emergency shut down scenario. The effective valve closure must be spread over multiple pipe periods to obtain a significant reduction of the peak pressure. Existing books on fluid transient provide more detail on efficient valve stroking (Tullis 1989; Streeter and Wylie 1993; Thorley 2004).

3.3.3. Emergency control procedures

Since WSS are spatially distributed, the power supply of valves and pumps in different parts of the system is delivered by a nearly-independent power supply. Therefore, local control systems may continue operating normally, after a power failure has occurred somewhere else in the network. The control systems may have a positive or negative effect on the propagation of hydraulic transients. The distributed nature of WSS and the presence of control systems may be exploited to counteract the negative effects of emergency scenarios.

If a centralised control system is available, valves may start closing or other pumps may ramp up as soon as a pump trip is detected. Even without a centralised control system, emergency control rules may be developed to detect power failures. These emergency control rules should be defined in such a way that false triggers are avoided during normal operations. An example of an emergency control rule is: *ESD valve closure is initiated if the discharge drops by more than 10% of the design discharge and the upstream pressure falls by at least 0.5 bar within 60 seconds.*

3.3.4. Anti-surge devices

The above-described measures may be combined with one or more of the following anti-surge devices in municipal water systems.

Devices, affecting velocity change in time	Pressure limiting devices
Surge vessel	By-pass check valve
Flywheel	Pressure relief valve
Surge tower	Combination air/vacuum valves
	Feed tank

Table 2. Summary of anti-surge devices

An important distinction is made in Table 2 between anti-surge devices that directly affect the rate of change in velocity and anti-surge devices that are activated at a certain condition. The anti-surge devices in the first category immediately affect the system response; they have an overall impact on system behaviour. The pressure-limiting devices generally have a local impact. Table 3 lists possible measures when certain performance criteria are violated.

The surge vessel is an effective (though relatively expensive) measure to protect the system downstream of the surge vessel against excessive transients. However, the hydraulic loads in the sub-system between suction tanks and the surge vessel will increase with the installation of a surge vessel. Special attention must be paid to the check valve requirements, because the fluid deceleration may lead to check valve slam and consequent damage. These local effects, caused by the installation of a surge vessel, should always be investigated in a detailed hydraulic model of the subsystem between tanks and surge vessels. This model may also reveal inadmissible pressures or anchor forces in the suction lines and headers, especially in systems with long suction lines (> 500 m). A sometimes-effective measure to reduce the local transients in the pumping station is to install the surge vessels at a certain distance from the pumping station.

Operation	Criterion Violation	Improvement
pump trip	low pressure	bypass pipe, flywheel
		larger pipe diameter
		air vessel, accumulator
		surge tower, surge vessel, feed tank
		air valve(s) at low pressure points in the system
		other pipe material with lower Young's modulus
pump trip	high pressure	air vessel with check valve and throttled by-pass
pump trip	reverse flow in pump	increase (check) valve closure rate by choosing an appropriate fast-closing check valve (e.g. nozzle type)
pump trip	rate of fluid deceleration through check valve (high pressure due to valve closure)	apply spring to reduce check valve closing time
		apply spring or counter weight with damper to increase check valve closing time and allow return flow
valve closure	high pressure (upstream)	air vessel
		slower valve closure
		pressure relief valve or damper at high pressure points
		higher pressure rating
valve closure	low pressure (downstream)	air vessel
		slower valve closure
		air valves at low pressure points
valve throttling	pressure instability	use multiple valves
		adjust control settings
drainage, filling	entrapped air	use air valves
		prevent drainage on shut-down

Table 3. Possible mitigating measures in case of violation of one or more performance criteria

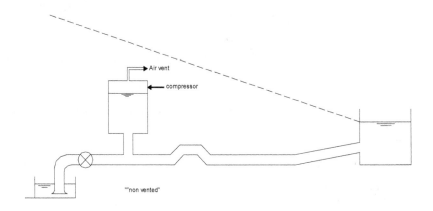

Figure 7. Non-aerated surge vessel

One of the disadvantages of a surge tower is its height (and thus cost and the siting challenges). If the capacity increases, so that the discharge head exceeds the surge tower level, then the surge tower cannot be used anymore. A surge tower is typically installed in the vicinity of a pumping station in order to protect the WSS downstream. A surge tower could also be installed upstream of a valve station to slow down the over pressure due to an emergency valve closure.

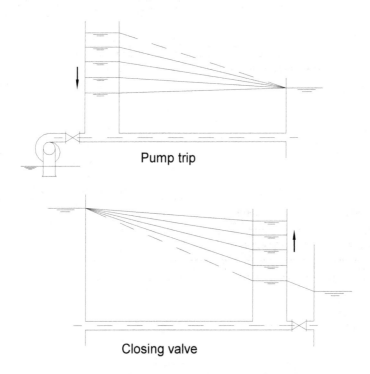

Figure 8. Surge tower near pumping station or valve station.

Another device that reduces the velocity change in time is the flywheel. A flywheel may be an effective measure for relatively short transmission lines connected to a tank farm or distribution network. A flywheel can be an attractive measure if the following conditions are met:

1. Pump speed variations are limited.

2. The pump motor can cope with the flywheel during pump start-up, which means that the motor is strong enough to accelerate the pump impeller - flywheel combination to the pump's rated speed. If the polar moment of pump and flywheel inertia is too large for the motor, then a motor-powered trip may occur and the rated speed cannot be reached.

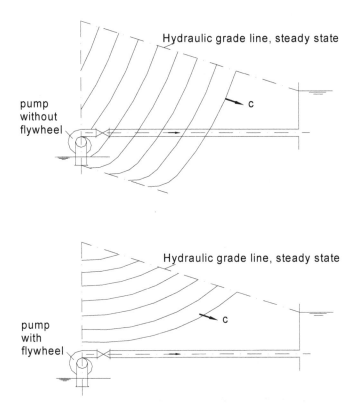

Figure 9. Effect of flywheel on transient pressure after power failure in the pumping station

A by-pass check valve is effective at sufficient suction pressure, which becomes available automatically in a booster station. Wavefront steepness is not affected until the by-pass check valve opens. A similar reasoning applies to the other pressure-limiting devices. Furthermore, the release of air pockets via air valves is an important source of inadmissible pressure shocks. Air release causes a velocity difference between the water columns on both sides of the air pocket. Upon release of the air pocket's last part, the velocity difference Δv must be balanced suddenly by creating a pressure shock of half the velocity difference (Figure 10). The magnitude of the pressure shock is computed by applying the Joukowsky law:

$$\Delta p = \pm \rho \cdot c \cdot \Delta v / 2 \qquad (5)$$

A large inflow capacity is generally positive to avoid vacuum conditions, but the outflow capacity of air valves must be designed with care.

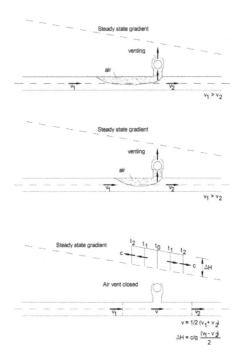

Figure 10. Pressure shock due to air valve slam.

3.4. Design of normal procedures and operational controls

The following scenarios may be considered as part of the normal operating procedures (see also appendix C.2.2. in standard NEN 3650-1:2012):

1. Start of pumping station in a primed system.

2. Normal stop of single pump or pumping station.

3. Commissioning tests.

4. Priming operation or pump start in partially primed system.

5. Procedure to drain (part of) the system for maintenance purposes.

6. Normal, scheduled, valve closure.

7. Stop of one pumping station or valve station and scheduled start of another source.

8. Other manipulations that result in acceleration or deceleration of the flow.

9. Switch-over procedures.

10. Risk assessment of resonance phenomena due to control loops.

Normal operating procedures should not trigger emergency controls. If this is the case, the control system or even the anti-surge devices may have to be modified. As a general rule for normal operations, discharge set-points in control systems tend to exaggerate transient events while pressure set-points automatically counteract the effect of transients. Two examples are given.

The first deals with a single pipeline used to fill a tank or supply reservoir. Suppose a downstream control valve is aiming for a certain discharge set-point to refill the tank or reservoir. If an upstream pump trip occurs, the control logic would lead to valve-opening in order to maintain the discharge set-point. This will lower the minimum pressures in the pipe system between the pumping station and the control valve. On the other hand, if the control valve aims for an upstream pressure set-point, the valve will immediately start closing as soon as the downsurge has arrived at the valve station, thereby counteracting the negative effect of the pump trip.

The second example is a distribution network in which four pumping stations need to maintain a certain network pressure. The pumping stations have independent power supply. Suppose that three pumping stations follow a demand prediction curve and the fourth pumping station is operating on a set-point for the network pressure. If a power failure occurs in one of the discharge-driven pumping stations, then the network pressure will drop initially. As a consequence the pump speed of the remaining two discharge-driven pumping stations will drop and the only pressure-driven pumping station will compensate temporarily not only the failing pumping station, but also the two other discharge-driven pumping stations. If all pumping stations would be pressure-driven pumping stations, then the failure of a single pumping station will cause all other pumping stations to increase their pump speed, so that the loss of one pumping stations is compensated by the three others.

The simulation of the normal operating procedures provides detailed knowledge on the dynamic behaviour of the WSS. This knowledge is useful during commissioning of the (modified) system. For example, a comparison of the simulated and measured pressure signals during commissioning may indicate whether the system is properly de-aerated.

It is emphasized that a simulation model is always a simplification of reality and simulation models should be used as a decision support tool, not as an exact predictor of reality. The design engineer of complex WSS must act like a devil's advocate in order to define scenarios that have a reasonable probability of occurrence and that may lead to extreme pressures or pressure gradients.

4. Modelling of water supply systems for transient analyses

This section provides some guidelines on the modelling of a pipeline system with respect to pressure surge calculations.

It is recommended to model the top of the pipes in computer models, because the dynamic behaviour may change significantly at low pressures due to gas release or cavitation.

The modelling and input uncertainties raise the question of which model parameter values should be applied in a particular simulation. The simulation results may be too optimistic if the model parameters are selected more or less arbitrarily. The model parameters should be selected such that the relevant output variables get their extreme values; this is called a conservative modelling approach. The conservative choice of input parameters is only possible in simple supply systems without active triggers for control procedures. Table 4 lists the parameter choice in the conservative modelling approach.

Critical Scenario	Output Criterion	Model Parameters (conservative approach)
any operation (cavitation not allowed)	max. pressure and min. pressure	high wave speed or low wave speed, high vapour pressure
upstream valve closure or pump trip (cavitation allowed from process requirements)	max. pressure due to cavity implosions	high vapour pressure
upstream valve closure or pump trip	min. pressure	high friction and low suction level
downstream valve closure	max. pressure	high friction and high suction level
upstream valve closure or pump trip (surge tower present)	min. pressure and min. surge tower level	low friction and low suction level
downstream valve closure (surge tower or present)	max. pressure, max. surge tower level	low friction and high suction level
critical operation	criterion	model parameters (conservative approach)
upstream valve closure or pump trip (air vessel present)	min. air vessel level	low friction and low suction level and isothermal air behaviour
upstream valve closure or pump trip (air vessel present)	min. pressure (close to air vessel)	low friction and low suction level and adiabatic air behaviour
upstream valve closure or pump trip (air vessel present)	min. pressure (downstream part)	high friction and low suction level and adiabatic air behaviour
downstream valve closure (air vessel present)	max. air vessel level	low friction and high suction level and isothermal air behaviour
downstream valve closure (air vessel present)	max. pressure (close to air vessel)	low friction and high suction level and adiabatic air behaviour
downstream valve closure (air vessel present)	max. pressure (upstream part)	high friction and high suction level and adiabatic air behaviour
Single pump trip, while others run	max. rate of fluid deceleration	high friction and low suction level

Table 4. Overview of conservative modelling parameters for certain critical scenarios and output criteria.

If control systems are triggered to counteract the negative effect of critical scenarios (pump trip, emergency shut down), then the extreme pressures may occur at other combinations of input parameters than listed in Table 4. Therefore, a sensitivity analysis or optimisation routine is strongly recommended to determine extreme pressures in these kind of complex water supply systems.

5. Concluding remarks

Since flow conditions inevitably change, pressure transient analysis is a fundamental part of WSS design and a careful analysis may contribute significantly to the reduction of water losses from these systems. It is shown that pressure transient analyses are indispensable in most stages of the life cycle of a water system. Section 2 shows that existing standards focus on a certain maximum allowable incidental pressure, but also emphasises that other evaluation criteria should be part of the surge analysis, including minimum pressures, component specific criteria and maximum allowable shock pressures. It is recommended that pressure shocks due to cavity collapse, air-release or undamped check valve closure should never exceed 50% of the design pressure. The main contributions of this paper, as compared to existing pressure transient design guidelines, include an overview of emergency scenarios and normal operating procedures to be considered, as well as the integrated design of control systems and anti-surge devices. These will lead to a safe, cost-effective, robust, energy-efficient and low-leaking water system.

Author details

Ivo Pothof[1,2*] and Bryan Karney[3]

*Address all correspondence to: ivo.pothof@deltares.nl

1 Deltares, MH Delft, The Netherlands

2 Delft University of Technology, Department of Water Management, Stevinweg, CN Delft, The Netherlands

3 University of Toronto, Canada and HydraTek and Associates Inc., Canada

References

[1] Boulos, P. F., B. W. Karney, et al. (2005). "Hydraulic transient guidelines for protecting water distribution systems." Journal / American Water Works Association 97(5): 111-124.

[2] Jung, B. S. and B. W. Karney (2009). "Systematic surge protection for worst-case tran-
    sient loadings in water distribution systems." Journal of Hydraulic Engineering
    135(3): 218-223.

[3] NEN (2012). Requirements for pipeline systems, Part 1 General. NEN, NEN.
    3650-1:2012.

[4] Pejovic, S. and A. P. Boldy (1992). "Guidelines to hydraulic transient analysis of
    pumping systems."

[5] Pothof, I. W. M. (1999). Review of standards and groud-rules on transients and leak
    detection. Computing and Control for the Water Industry. Exeter, RSP Ltd, England.

[6] Pothof, I. W. M. and G. McNulty (2001). Ground-rules proposal on pressure transi-
    ents. Computing and Control for the Water Industry. Leicester, RSP Ltd, England.

[7] Streeter, V. L. and E. B. Wylie (1993). Fluid transients in systems. New York, Pren-
    tice-Hall.

[8] Thorley, A. R. D. (2004). Fluid Transients in Pipeline Systems. London, UK, Profes-
    sional Engineering Publishing Ltd.

[9] Tullis, J. P. (1989). Hydraulics of pipelines, pumps, valves, cavitation, transients.
    New York, John Wiley & Sons.

Chapter 3

Model Based Sustainable Management of Regional Water Supply Systems

Thomas Bernard, Oliver Krol,
Thomas Rauschenbach and Divas Karimanzira

Additional information is available at the end of the chapter

1. Introduction

The sustainable management of the water resources and a safe supply of drinking water will play a key role for the development of the human prosperity in the following decades. The fast growth of many cities puts a large pressure on the local water resources, especially in regions with arid or semi-arid climate. A research project at Fraunhofer IOSB and AST has aimed to investigate ways for economic and sustainable use of the available water resources in the region of the capital of China, Beijing [1].

A main issue of the project is to develop components for a model-based decision support system (DSS), which will assist the local water authority in management, maintenance and extension of the water supply system at hand [2]. This paper deals with the derivation of suitable management strategies for a mid (till long) term horizon based on assumptions for future environmental and socio-economic conditions, which are provided by other modules of the DSS. The general structure of the proposed optimal control DSS is shown in Fig. 1.

An overview about several DSS concepts and implementations is provided in [3, 4]. A challenge of the given problem is the large area of the water supply system. The water management has to consider the total water resources of five river basins with an area of 16,800 km² as well as large groundwater storage in the plain with an area of 6,300 km². The main portion of the annual precipitation (85%) in this semi-arid region is falling from June to September leading to a highly uneven distribution throughout the year. The formerly abundant groundwater resources have been overexploited over the last decades resulting in a strong decline of the groundwater head (up to 40 m). Five reservoirs are important for the management of the surface water in the considered area, where the two largest account for about

90% of the total storage capacity of roughly m³. The water is distributed to the customers using rivers and artificial transport ways (channels, pipes) of a total length of about 400 km.

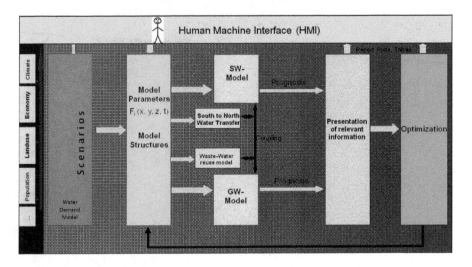

Figure 1. Structure of the proposed decision support system (DSS) for the region of Beijing

A common approach for policy generation in this field of application is to use mathematical programming techniques based on a dynamic model of the essential elements of the water allocation and distribution system [3]. The great impact of the groundwater storage for the supply system at hand requires a more detailed description of the groundwater flow dynamics compared to other known DSS implementations. Therefore a 3D Finite-Element model of the plain region has been developed. However, a direct integration of this 3D model into an optimal control framework is not possible due to its computational costs. A trajectory based model reduction scheme is proposed, which guarantees a very fast response of the DSS in combination with a specially tailored non-linear programming algorithm.

The chapter is organized as follows: In section 2 the essential models (surface and groundwater models) and the model reduction approach are described. While the formulation of the optimal control problem is subject of section 3, the numerical solution of the large scale structured non-linear programming problem is described in section 4. First results of the optimal water management approach are presented in section 5.

2. Water allocation model

The water allocation model can be divided into the surface water model and the groundwater model. The parts of the water allocation model are described in the sequel.

2.1. Surface water model

The surface water model has to comprise all important elements for the allocation, storage and distribution of water within the considered region. The intended field of application of the decision support system under development embraces management and upgrading strategies for the mid and long term range. This implies a significant simplification for the process model, because the retention time along the different transport elements (river reaches, channel or pipelines) is less than the desired minimum time step for decision of one day. Therefore, a simple static approach for flow processes is sufficient and the use of so-phisticated models for the dynamics of wave propagation (like e. g. Saint-Venant-Equations) with respect to control decisions is avoided. In this case, the flow characteristics are repre-sented by simple lag elements of the first order combined with dead-time elements. Given y and u as an output and input, respectively. The following mathematical relationship indi-cates a lag element of the first order.

$$T_1 \cdot \frac{dy}{dt} + y = u \tag{1}$$

where T_1 is the only parameter and indicates the time lag; t indicates the time. The following relationship represents the dead-time element:

$$y = u(t - T_t) \tag{2}$$

In this case the parameter is the dead-time T_t, which is thus the measure of the time taken for water to flow in a conduit over a known distance.

The surface water system is described as a directed graph. The edges characterize the trans-port elements and introduce only a variable for the discharge. The nodes represent reser-voirs, lakes, points of water supply or extraction and simple junction points. Every node constitutes a balance equation involving the edges linked with and possibly the storage vol-ume. The sole nonlinearity results from modeling the evaporation from the water surface of the storages (volume-area-curve), which is described by a piecewise polynomial approach. At time step k, the volume of a reservoir node evolves as follows:

$$V_j^{k+1} = V_j^k + \Delta t^k \left(\sum_{i \in E(j)} Q_i^k + A_{O,j}^k q_{evpot,j}^k - q_{seep,j}^k + A_{O\max} q_{prec,j}^k \right), \tag{3}$$

where $A_{O,j}^k = f(V_j^k)$, the storage volume is denoted by V, the discharge into and from the storage is denoted by Q and discrete time step is denoted by Δt. The total evaporation from the reservoir depends on the water surface A_0 and the potential evaporation q_{evpot}. q_{seep} de-notes the seepage from the reservoir to the groundwater and q_{prec} specifies the precipitation.

Channels with a very low slope are modeled as water storage. The level dependent upper bound for the channel outflow is derived from a steady state level-flow relation like e.g. Chezy-Manning friction formula and is directly added as constraint to the optimization problem.

The Structure of the Beijing water supply system is shown in Fig. 2. Firstly, there are the four main reservoirs Miyun, Huairou, Baihebao and Guanting. Further sources are groundwater storages and water transfers. Secondly, there are the water transportation systems such as channels and rivers. Miyun reservoir and Huairou reservoir are connected to Beijing by the Miyun-Beijing water diversion. In the simulation model the arrows describe hydraulic behavior of water flow. Baihebao and Guanting reservoir are connected by tunnel and river Guishui. From Guanting water runs into the Yongding river water diversion system to Beijing. Existing retention areas for flood control are also considered in the simulation model.

The surface water from channels and rivers is delivered to the customers in different ways, directly or through the surface waterworks. Groundwater is distributed to the customers through ground waterworks as well as motor-pumped wells. Therefore, the waterworks build the third part of the Beijing water supply system. The last category is made up of all the customer groups (agriculture, industry, households and environment). To complete up the cycle, catchments area models are integrated in the system to take into account precipitation and evapotranspiration.

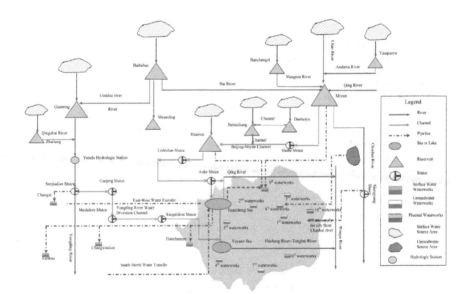

Figure 2. Structure of the Beijing water supply system

The surface water simulation model has been implemented in Matlab/Simulink using the toolbox "WaterLib" [5] and contains the most important elements of the drinking water supply system of Beijing. The simulation model was developed to reach a sufficient accuracy as well as a high simulation speed. A one-year simulation is carried out in a simulation time of several seconds. This is a very important condition for using the model in the decision support system.

2.2. Groundwater water model

The most important water resource in the considered area is groundwater that is modeled by a dynamic spatially distributed finite element groundwater model. The governing equation for groundwater flow is Darcy's law [6] describing slow streams through unconfined aquifers. Combining Darcy's law with mass conservation yields the partial differential equation (4) which is a diffusion equation.

$$S_0 \bar{h} - \nabla \cdot \left(k_f \nabla h \right) = Q_{rech} - Q_{expl} \qquad (4)$$

In (4) denotes h the hydraulic head (which corresponds to the groundwater level) and k_f the hydraulic conductivity that governs the hydrogeological properties of the soil. S_0 denotes the specific storage coefficient. The terms on the right hand side of (4) summarize all sources and sinks that coincide with the time dependent groundwater exploitation due to industry, households and agriculture (Q_{expl}) and recharge e. g. due to precipitation and irrigation (Q_{rech}) in Ω.

The partial differential equation (4) is an initial-boundary value problem which has to be solved numerically for h in the 3 dimensional model domain Ω. The groundwater model has been implemented using FEFLOW, which is a Finite Element (FEM) software specialized on subsurface flow [7]. The initial condition is h (Ω, t_0) (groundwater surface) at the initial time t0. The inflow/outflow is described by Dirichlet boundary conditions, i.e. h ($\partial\Omega$) at the boundary $\partial\Omega$ and by well boundary conditions, that define a particular volume rate into or out of Ω. The advantage of the latter one is that they are scalable. The 3D FEM model consists of more than 150,000 nodes, distributed on 25 layers (cf. Fig. 3). Huge computational costs result from this high resolution. The simulation of 5 years needs ~15 Minutes on an Intel Core 2 Duo CPU (2.5 GHz). Hence it is very time consuming to calculate optimal water allocation strategies with the 3D FEM groundwater model. This is the motivation for model reduction (see subsection 2.3).

The main task with respect to the groundwater model is the parameterization of the large-scaled model covering an area of 6,300 km². On the one hand the time independent soil parameters k_f, S_0 have to be estimated and generalized for the whole domain Ω by a (small) set of measured values. On the other hand the source / sink terms Q_{expl}, Q_{rech} have to be calculated time dependent. For these calculations time dependent maps of precipitation and water demand are needed. The water demand is splitted into the three user groups households, industry and agriculture (see [8] for details). This parameterization issue is supported by powerful geographical information systems (GIS).

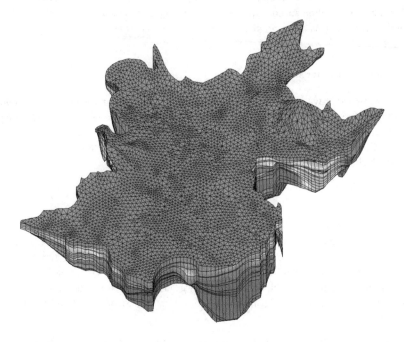

Figure 3. Mesh of the 3D Finite Element groundwater model of the region of Beijing

2.2.1. Hydrogeological conditions and derivation of hydrogeological parameters

The groundwater model area is located at the northern part of the North China Plain (NCP), which is the largest alluvial plain of eastern Asia. The NCP is a basin with quaternary aged surficial deposits (loess, sand, gravel and boulder, silt and clay). According to the hydrogeo-logical profiles of Beijing the quaternary system in this region is fairly complicated. A great variety of different sedimentary facies exists with different thicknesses ranging from several tens of meters around the piedmont area to 150 - 350 meters in the northern central part of the NCP [9]. Groundwater is exploited in the layers of quaternary deposits, i.e. in the loose stratum/porous aquifers with high to very high water storage capacities. From the Taihang Mountains in the west to east there are two main geomorphological units in the model area: the piedmont plain below the mountain escarpments and the flood plain. In the piedmont plain the aquifers structure is coarse and becomes finer from west to east. In the flood plain the structure of aquifer is fine with silt sand, clay and silt interlay and in areas of ancient rivers and paleochannels the aquifer is composed mainly of gravels and coarse sands with good permeability. Therefore the distribution of groundwater in the Beijing region is inhomogeneous. Regions of high abundance and high yielding porous groundwater aquifers are the piedmont plains and the northeastern districts of Miyun, Huairou and Shunyi whereas less yielding aquifers are found in the Yangqing and Tong districts. In the transition zone from the Taihang Mountains to the NCP the quarternary sediments with low thickness of e. g. some tens of meters are lying on the older rock formations of the regions.

In the mountainous districts unstable groundwater distributions were assumed in dependence on the form of the rocks with geological discontinuities (fractures, joints, dissolution features) and the groundwater flow. In the transition area from the Taihang and Yanshan Mountains to the NCP stratigraphic sequences of various ages ranging from archaean metamorphic rocks to quaternary are documented in the geological and hydrogeological maps. A detailed description of the geological and hydrogeological conditions can be found in [10].

On the base of a conceptual geological model and a structured horizontal (2D) groundwater model, a horizontal and vertical structured 3D - groundwater model was developed describing the saturated zone till approx. 200 m depth below ground surface (bgs.) in the area of the quaternary sediments of the NCP. In addition the borehole data from approx. 125 drillings situated in the model area were used in the groundwater model. Although a quite homogeneous distribution of the boreholes was given, one measurement point represents an area of about 50 km^2 which is only a rare database for modelling subsurface conditions.

There can be found strong variations in structure and thickness of the loose stratum sediments in the model area. The evaluation of all data (borehole data, geological and hydrogeological maps and profiles, ground water levels from observation wells, literature etc.) shows that the large number of the water bearing layers can be summarised in up to three essential ground water aquifers according to present knowledge on regional level. These aquifer systems are from top to down:

• Aquifer I: Shallow aquifer in approx. 5 m to 30 m depth bgs.

• Aquifer II: Primary aquifer, till approx. 120 m depth bgs.

• Aquifer III in the depth area of approx. 120/140 m to 200/260 m bgs

The aquifers are separated by less permeable layers or aquitards, above all fine sands, silts and clays. It can be assumed that the three essential aquifers are not completely independent from each other, i.e. a groundwater exchange takes place between them in a certain range. Where low permeable layers or aquitards are absent or have a low thickness two aquifers can form a hydraulic unity as in the area of piedmont plains. Thus in the piedmont plains only one porous aquifer between the unsaturated loess top set layers and the bedrock was assumed. In regions, in which the separating layers have bigger thickness and larger extension, local confined aquifers can appear. Because of morphology and evolution processes perched aquifers can appear within the loess deposits. All these local effects are summarised in the above mentioned three essential aquifers.

The piedmont areas of the Taihang Mountains and the Yanshan Mountains along the western boundary and the northern/northeastern boundary of the area are the areas where groundwater inflow into the plains contributes to the groundwater recharge of the confined and unconfined aquifers. Because of the multi-layered geological structure of the loose stratum in the model area, consisting of loess, alluvial loess, sand-gravel-cobble-boulder sediments with more or less mighty clay and silt inclusions, the details of hydro-geological conditions are complicated. Therefore the used spatial distribution of the hydraulic conductivity and specific yield (both important for good model results) are adequate phenomenological descriptions of mean values and not a detailed representation of local conditions in reality.

The aquifer characteristics were determined mainly on the base of the interpretation and evaluation of the above mentioned borehole data. The borehole data represent the geological layers (boring logs) at single points. They show high variations from one point to another. In particular the hydrogeological parameters k_f and S0 were deduced from these borehole data due to the following procedure:

1.   In a first step the local single point data have to be transformed in values representative for an associated area (meso-scale values) [11]. For this step the information and data from thematic maps (e.g. Beijing hydrogeological maps lithology map, water abundance map, etc.) were included. The validity range of these meso-scale values could be specified with the informations from a water abundance map representing the water yield and water storage capability. These data resulted primarily on measurements in water exploitation wells. With this approach meso-scale values (for k_f -and S0) and their spatial distribution could be determined.

2.   In a second step the discrete meso-scale values were interpolated within the model area

3.   In a third step the absolute values of the smoothed spatial distribution of k_f and S0 were adapted to fulfil the water budget of the model area.

On the basis of the derived values a fine tuning was realized in order to get minimal differences between calculated and measured groundwater surface map.

The k_f -values of the aquifers range from 2.0 10^{-3}m/s in the region of the piedmont plains and the alluvial fan plains to 0.1 10^{-4}m/s in the flood plains. The loose stratum depositions can be classified according to German Institute for Standardization Guideline 18130 as permeable to very permeable.

In [12] a mean storativity in the range of 0.08 and 0.18 has been estimated. Due to hydrogeological investigations of borehole data a regionalisation of the hydrogeological parameters could be performed, yielding a mean storativity of 0.13. The values of S0 changes from 0.19 (piedmont plains) to 0.03 (flood plains). This spatial distribution of k_f - and S_0-values was also a basis for the regionalisation of the inflow boundary conditions.

Inflow conditions

The inflow from the north and west into the model area is governed by the transition zone between the mountain terrain and the plain where a high conductivity can be assumed. Here the inflow consists of the surface water run o_ from the mountains that depends on the precipitation rate. But due to investigations even after a number of dry years the total inflow did not disappear, but decreased from a long term mean value of about 0.7 10^9m³/a to 0.4 10^9 m³/a could be observed by the Chinese partners. This amount of water could not only result from rainfall runoff from the mountainous domain. Therefore it was assumed a split of the horizontal groundwater recharge from groundwater inflow into the model area (see Fig. 4). One part is a rain-dependent contribution which in 'normal years', with a mean precipitation rate of about 590 mm/a is in the range of 0.4 10^9 m³/a. This value is scaled in dependency of the mean precipitation rate of the current year. In a dry year with a precipitation rate of 380 mm/a, for instance, the rain dependent inflow to 0.26 10^9 m³/a. For this contribution the imagination is that a part of the precipitation of the mountain slopes infiltrate on the surface and percolate down to the bedrock basis. On the relatively impermeable bedrock the water flows as shallow groundwater aquifer in the loose stratum with low thickness into the model area. The loose stratum depositions in the piedmont plains consist of boulder, gravels and sand with inclusions of local loess and loessloam lenses. These loose stratum depositions are well permeable and this inflow contribution from the mountains is asssumed to be directly dependent on the precipitation.

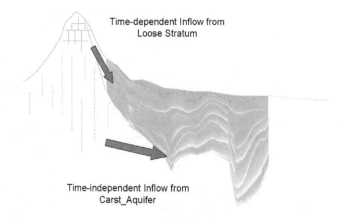

Time-dependent Inflow from
Loose Stratum

Time-independent Inflow from
Carst_Aquifer

Figure 4. The groundwater inflow regime into the NCP

The second contribution to the horizontal groundwater recharge from the mountainous area is of about 0.3 10^9 m³/a and it was implemented deeper. Here the underlying idea is that this component corresponds to the part of the precipitation which infiltrates in the mountainous area through clefts into the deeper rock formations. The groundwater flow system in the bedrock consists of macroscopic structures and cavities linked with each other, as for example fracture networks, faults, layer joints, dissolution features and conduits etc. The groundwater inflow depths were assumed according to the distribution of the water bearing carbonate rocks, sandstones and crystalline rocks in the hydrogeological map of Beijing. These so-called deep inflows are not dependent directly on the precipitation and change only in terms decades.

The total horizontal groundwater recharge (inflow) ranges from 0.55 to 0.75 10^9 m³/a. The quantitative split is to some extend arbitrary and based only on plausibility considerations since none of the components can be measured directly.

2.2.2. Derivation of timedependent groundwater recharge and exploitation data

In order to determine the time-dependent groundwater recharge and exploitation we follow the subsequent procedure:

1. It starts with stating a long term mean value budget for the considered area for getting an idea which fluxes are in which order.

2. In a second step the budget data are regionalized by means of land use maps.

3. Finally the temporal distribution is taken into account by implementing the crop water need (agricultural water demand) and precipitation as dynamic input data such that in the end due to balancing all data, e.g. irrigation, evapotranspiration, etc, become time dependent quantities.

The first task in setting up models covering the water resources of an area of this size is to construct a water budget. The water budget is a theoretical device that supports structuring the water resource system and identifying the most important water fluxes. Here fluxes into and out of the system has to be collected as well as the water fluxes within the model area. The intension must be to realize the relation of water fluxes to each other, to quantify them, separating the more important from negligible water fluxes and to estimate the error that happens due to neglecting them. Since most of the quantities in the water budget are not independent from each other, the quantification of the water budget must be an iterative process. Fig. 5 illustrates the water budget of the region of Beijing. All the before mentioned fluxes are entried. The width of the arrows corresponds to the quantity of the fluxes.

Since the groundwater model is spatially distributed the input data for the groundwater recharge and the exploitation have to be spatially distributed as well. The above mentioned water budget can be regarded as a lumped parameter model for the whole region. The next step is to relate these data to locations by adapting the water balance with respect to specific regions.

• An approach which is followed quite often is to regionalize by means of land use maps. The land use of the considered region is depicted in Fig. 6 showing eleven land use classes. These classes can be summed up to the following four classes:urban and paved areas,

• agriculturally used and irrigated areas,

• water areas and

• non-cultivated areas

Water Budget

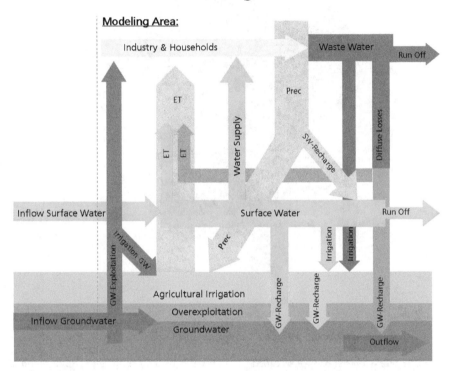

Figure 5. Water budget of the region of Beijing

Regionalization of the water budget means to adapt and evaluate the balance equation

$$P = GWR + ET + SWR \tag{5}$$

for each of these classes. It denotes P the precipitation rate, GWR is the groundwater recharge, ET the evapotranspiration SWR corresponds to the surface water runoff. There are some quantities which are relevant for every class, like infiltration and there are some which are specific for a certain group. And for other classes additional quantities have to be added, i.e. irrigation rates (IRR) for agriculturally used areas.

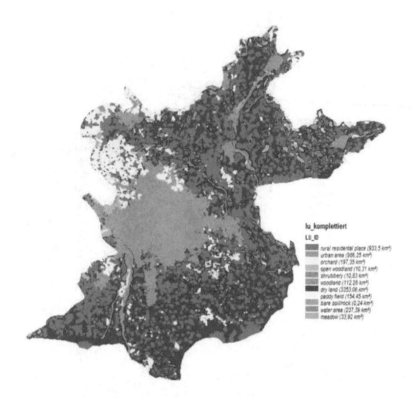

Figure 6. Landuse map of the region of Beijing

2.2.3. Software realisation of the parameter calculation within the DSS

The DSS works on the base of so-called *scenarios* whereby a scenario can be understood as set of input data that represents either a real historical situation or an imaginable situation in future with respect to precipitation, exploitation, water use etc. This complete set of input data is applied to the hydrological model to obtain an answer to the question 'What are the effects to the water allocation system'. The historical scenarios usually are used for the validation of the models whereas future scenarios should deliver guidelines for the best way to act in future under certain circumstances. In order to obtain reasonable results from the simulation the scenario has to be complete and consistent. Otherwise the models will gain unreasonable results. A scenario is complete if all data are available the hydrological model expects. In order to keep the system stable often default values are inserted if no data are assigned. But in this case the question is if the data are consistent.

The consistency of data can only be assured by the user. So if we use precipitation rates for groundwater model and for the surface water model for instance in general the information is provided in different data set and formats. In this case the user is responsible, that the data for the groundwater model match the data for the surface water model, i.e. that at same time periods in same regions the same precipitation rates are considered.

Nevertheless, the user can be supported by a parameterization software, as it was realized within the Scenario Wizard of the Beijing DSS to ensure consistency as good as possible. With respect to the groundwater model the scenario is complete and consistent only if the before mentioned data are available for the complete simulation horizon:

- Initial conditions

- Boundary conditions (well fields and inflow)

- Spatially distributed groundwater recharge

- Spatially distributed exploitation

In order to obtain a complete and consistent set of input data for the groundwater model the user has to pass through the groundwater panel of the DSS Scenario Wizard and a so-called groundwater project is created. A groundwater project is a part of the scenario containing all groundwater relevant data.

Within the GW panel the user has to execute five subpanels such that in the end at least a complete set of input data for the groundwater model is generated.

1. The first subpanel calculates the surface water recharge. For this a weighting map is required that determines how much precipitation becomes surface water in direct or indirect manner. On the other hand a reliable precipitation map for the entire model area is necessary. Since the precipitation is not constant over the year we also need temporal weights that define the temporal distribution of the precipitation rate. The Scenario wizard provides a number of precipitation maps of the past which can be also used for future scenarios. The surface water recharge is not a direct input data for the groundwater model but it is required to determine the distribution of exploitation and groundwater recharge in time and space.

2. In the second step the agricultural irrigation from groundwater is and the total spatially distributed exploitation from groundwater is derived. For these calculations a map of the agricultural used areas (irrigated areas) is needed as well as corresponding information about the crop water need/ agricultural water demand. In addition some information about the irrigation from surface water and waste water are requested. Since the agricultural water demand is not constant during the year a temporal distribution is needed as well.

3. In the third subpanel the minimal surface water inflow and the diffuse losses in urban areas is computed and therefore maps of water areas and urban areas are requested. The resulting information is incorporated into the calculation of the groundwater recharge from surface water areas and from urban areas.

4. In this subpanel all input data with respect to the well fields and the inflow parameters have to be defined and entered in to table which asks for yearly data of the ex-

ploitation rates from the different well fields and the upper and deeper inflow into the model area.

5. In the last subpanel the generation of the spatial and temporal distribution of the groundwater recharge data is performed. Here the user again has to provide a weighting map that determines how much of the supplied water becomes groundwater recharge and how much becomes evapotranspiration. The rest of the required data was already determined in the before described subpanels. This step finishes the creation of the groundwater project and all the generated data are documented in a specific groundwater scenario report which is saved within the corresponding project directory. After the simulation of a scenario with a groundwater simulation run a report about simulation results is saved in the project directory as well.

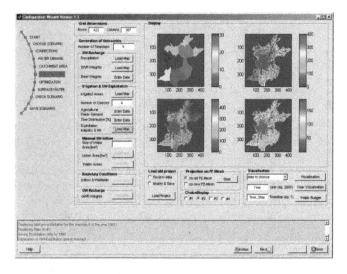

Figure 7. Graphical user interface for the generation of input data like groundwater recharge maps

All computed data are saved as ASCII grid maps which can be read by any GIS system. Figure 7 shows the graphical user interface generating groundwater maps.

In order to gain the project data accessible to the FEFLOW simulation model the spatially distributed data have to be projected on the finite element mesh. But in this step no new information is gained therefore it is not a procedure which is important for the scenario generation but it is only a technical requirement.

2.3. Derivation of reduced groundwater model

For the optimization of the water allocation system the full 3D FEM model (with > 150,000 nodes) is not very suited due to the mentioned large computational time. As for the optimization task a prediction of the hydraulic head (groundwater level) at a set of representative and fixed points is sufficient, an input-output model (e.g. a linear state space model) with considerably smaller order n (e.g. n < 50) than the original FEM model has to be derived.

Methods for model reduction of those large scale systems have gained increasing importance in the last few years [13]. Two main classes of methods for model reduction can be identified, namely methods based on singular value decomposition (SVD) and Krylov based methods. SVD based methods are suited for linear systems and nonlinear systems of an order n < 500 (e. g. balanced truncation for linear systems, proper orthogonal decomposition (POD) for nonlinear systems). Most of these methods have favourable properties like global error bounds and preservation of stability [13]. Krylov based methods are numerically very efficient as only matrix multiplications and no matrix factorization or inversion are needed. Hence they are also suited for large-scale systems. Unfortunately, global error bounds and preservation of stability cannot be guaranteed. Hence actual research is focused on the development of concepts which combine elements of SVD and Krylov based methods [13].

All of these approaches have in common that they aim to approximate the state vector x with respect to a performance criterion, e. g. minimize the deviation between original system and reduced system for a given test input. As a black box input-output model would be sufficient for our purposes there is no need to approximate the whole state vector x. Furthermore, the dimension n of a reduced model which approximates the whole state space vector x would be in most cases n > 100. With this dimension, for the given optimization problem the solution time would be unacceptably high (~ hours). Last but not least the use of the commercial software FEFLOW also prevents the application of e. g. a Krylov based method as no model representation (e. g. state space model) is provided by the software.

The basic idea of the proposed model reduction method is sketched in Fig. 8. We assume the existence of a reference scenario which means that the time dependent input parameters uref(t) of the FEM groundwater model (especially groundwater exploitation Qexpl and recharge Qrech) are determined for the whole optimization horizon. In practical cases these reference scenarios are mostly available or can be generated by plausible assumptions. Hence the task consists in the derivation of a model which approximates the behavior of the full FEM model in the case that the input parameters u differ from uref(t). This model is gained by identification techniques: Test signals (e.g. steps) are added to the reference input uref(t) (dimension p) and the corresponding deviations from the reference output yref(t) (dimension q) are identified. Doing this separately for every component of the input-/output vectors u and y, we finally merge the (p q) single input-single output (SISO) models to a multi input-multi output (MIMO) model. For the groundwater model, the input parameters are e. g. cumulated (e.g. spatially integrated) exploitation of certain regions or cumulated exploitations of large well fields. The output parameters of the groundwater model are the hydraulic head at representative points ("observation wells"). In our application 13 input and 13 output parameters have been defined by the users: The inputs consist by 9 counties and 4 wellfields, the 13 output parameters are 12 observation wells and the mean hydraulic head of the whole area of the water supply system. As the slow stream groundwater flow can be interpreted as diffusion process (cf. equation (4)) only nearby located input and output parameters (e.g. regions/wellfields and the corresponding observation wells) have some correlation and a SISO model with these input-/output combinations can be gained. Due to this physical reason the number of relevant SISO models is relatively small and hence the resulting MIMO model of relatively low dimension (n < 50) which is appropriate for the optimization problem. This proposed approach can be called trajectory and identification based model reduction. A similar approach (for a nonlinear large scale system) is found in [14].

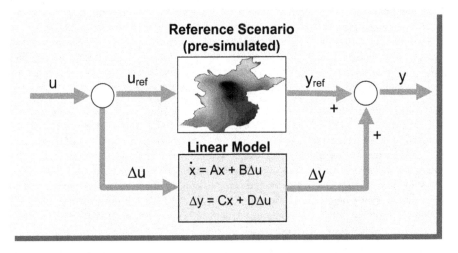

Figure 8. Reduced groundwater model as a linear state space model in combination with a pre-simulated reference scenario.

3. Water resources management as optimal control problem

The water resources allocation problem is formulated as a discrete-time optimal control problem:

$$\min_{\mathbf{u}^k,\, k=1,\ldots,K} \left\{ F\left(\mathbf{x}^K\right) + \sum_{k=0}^{K-1} \mathbf{f}_0^k\left(\mathbf{x}^k, \mathbf{u}^k, \mathbf{z}^k\right) \right\} \tag{6}$$

subject to

$$\mathbf{x}^0 = \mathbf{x}\left(t_0\right) \tag{7}$$

$$\mathbf{x}^{k+1} = \mathbf{f}^k\left(\mathbf{x}^k, \mathbf{u}^k, \mathbf{z}^k\right) \tag{8}$$

$$\mathbf{h}^k\left(\mathbf{x}^k, \mathbf{u}^k, \mathbf{z}^k\right) = \mathbf{0} \tag{9}$$

$$\mathbf{g}^k\left(\mathbf{x}^k, \mathbf{u}^k, \mathbf{z}^k\right) \leq \mathbf{0} \tag{10}$$

The state variables x are the volume content of the reservoirs and channels with small slope and the states of the reduced groundwater model. Control variables u are the discharge of the transport elements as well as the water demand of the customers. The uncontrollable inputs z are the direct precipitation and the potential evaporation for the reservoirs and the flow of rivers entering the considered region, which is derived by means of rainfall-runoff-models. The number of time steps within the optimization horizon is denoted by K.

The process equations (8) consist of the balance equations of the storage nodes and the reduced groundwater model. The balance equations of the non-storage nodes are formulated as general equality constraints (9). The objective function (6) contains the goals of the water management, which are primarily the fulfillment of the customer demand, the compliance with targets for the reservoir and groundwater storage volume and the delivery of water with respect to environmental purposes. Therefore quadratic terms are formulated, which penalize the deviations from desired values, like e.g. for the demand deficit of the demand node j:

$$f_0^k\left(u_j^k\right) = \rho_j^k \Delta t^k \frac{\left(u_j^k - q_{dem,ref,j}^k\right)^2}{q_{dem,ref,j}^{k\,2}} \tag{11}$$

where $q_{dem,ref,j}$ is the demand of and u_j is the discharge delivered to the customer. While this term applies for every time step within the optimization horizon, other terms are formulated only for the final point of the horizon, like e.g. for the desired volume content of the reservoirs.

The inequality constraints (10) follow from the technical capabilities of the water distribution system and rules to guarantee safe operation, which are simple bounds for the control variables:

$$\mathbf{u}_{min}^k \leq \mathbf{u}^k \leq \mathbf{u}_{max}^k \tag{12}$$

as well as constraints for the reservoir volume x_v:

$$\mathbf{x}^k_{v,\min} \leq \mathbf{x}^k_v \leq \mathbf{x}^k_{v,\max} \tag{13}$$

and the hydraulic head h_{hydr} of the observation wells:

$$\mathbf{h}^k_{hydr,\min} \leq g^k\left(\mathbf{x}^k_{gw}\right) \leq \mathbf{h}^k_{hydr,\max} \tag{14}$$

With respect to the practical applicability selected parts of the inequality constraints (10) can be relaxed in order to avoid infeasible optimization problem with respect to unrealistic management demands.

4. Numerical solution of the optimal control problem

The optimal control problem is numerically solved as large scale structured non-linear programming problem (NLP):

$$\min_{\mathbf{y}} \left\{ J(\mathbf{y}) \mid \mathbf{h}(\mathbf{y}) = 0; \mathbf{g}(\mathbf{y}) \leq 0 \right\} \tag{15}$$

The optimization variables are the state and control variables of the several stages in time:

$$\mathbf{y} = \left[\left(\mathbf{x}^0\right)^{\mathrm{T}} \left(\mathbf{u}^0\right)^{\mathrm{T}} \cdots \left(\mathbf{x}^{K-1}\right)^{\mathrm{T}} \left(\mathbf{u}^{K-1}\right)^{\mathrm{T}} \left(\mathbf{x}^K\right)^{\mathrm{T}} \right]^{\mathrm{T}} \tag{16}$$

The state equations of the process model are incorporated as equality constraints:

$$\mathbf{h}(\mathbf{y}) = \begin{bmatrix} \mathbf{x}^0 - \mathbf{x}(t_0) \\ \mathbf{f}^0\left(\mathbf{x}^0, \mathbf{u}^0, \mathbf{z}^0\right) - \mathbf{x}^1 \\ \vdots \\ \mathbf{f}^{K-1}\left(\mathbf{x}^{K-1}, \mathbf{u}^{K-1}, \mathbf{z}^{K-1}\right) - \mathbf{x}^K \end{bmatrix} \tag{17}$$

The advantage of this problem formulation with $(K(n+m)+n)$ optimization variables is the special sparsity structure with a block-diagonal Hessian-matrix and block-banded Jacobian matrices (k: number of time steps, n: number of state variables, m: number of control variables), which follows from the fact, that the process equations as coupling element between adjusting stages are linear in the state variables x^{k+1}. The numerical solution of this problem requires about $(K(n+m)^3)$ basic arithmetic operations.

One alternative approach consists of eliminating the state variables from the optimization problem. The reduced dimension of the according non-linear programming problem with (K m) optimization variables comes along with a loss of structure. The Hessian and Jacobian matrices are full. Because of the solution effort of order (K m)^3 and the large amount of control variables (number of edges in the network description) this approach is not promising for this special field of application.

For the numerical solution of large scale non-linear programming problems interior point (IP) solver have become popular during the last years because of their superior behavior for NLPs with many inequality constraints. In this approach the objective function is expanded by adding barrier terms for the inequality constraints:

$$\min_{y} \left\{ J(y) + \mu \sum_{j=1}^{n_g} \ln\left(-g_j(y)\right) \mid h(y) = 0 \right\} \tag{18}$$

The solution of the original NLP (15) results from the subsequent solution of (17) with a decaying sequence of $\mu \rightarrow 0$. The identification of right active set with its combinatorial complexity is avoided. The state of the art non-linear interior point solver IPOPT is used for the application at hand [15]. The interface for multistage optimal control problems of the optimization solver HQP [16], which provides an efficient way for problem formulation along with routines for the derivation of ∇J, ∇h, ∇g by means of automatic differentiation [17], is used and coupled to IPOPT.

A typical water management problem (horizon of five years, discretization of one month) has about 8000 optimization variables, 5500 equality constraints and 7200 inequality constraints. The numerical solution takes approximately 60 iterations and a calculation time of 10 seconds on an Intel Core 2 Duo CPU (2.5 GHz). Table 1 shows the linear dependency of the numerical solution effort from the number of time steps within the optimization horizon.

optimization horizon	Numerical solution effort		
optimization variables	Iterations	main storage [MB]	calculation time [s]
30days	7898	56	52	9.6
10 days	23880	66	145	35.1
5 days	47853	56	286	61.7

Table 1. Numerical solution effort in dependence on the optimization horizon

5. First results of the optimal water management approach

The proposed concept for optimal water management is applied to the Beijing region. The region has precipitation, which varies geographically, seasonally and yearly. Eighty-five percent of rainfall falls between July and September. Groundwater is the most important source of water for the Beijing region, covers about 50-70%. Beijing has suffered from over exploitation of this source over the years. Surface water supply in the Beijing region depends mainly on upstream inflows of the major river systems Chaobai, North Grand Canal and Yongding. Aside from problems such as excessive withdrawal and water quality deterioration of surface waters, the lack of regional coordination leads to issues such as uncoordinated withdrawals. Besides these problems, the water supply system is subject to other common problems, such as rapid population growth and urbanization, decentralized reservoir/groundwater management, changing attitude towards sustainability and attribution to greater attention of environmental issues.

The optimal water management system is evaluated for three scenarios based on the same set of input data, which are a combination of historical rainfall measurements and customer demands reflecting the predicted development of population size and economic growth. The objective function contains four quadratic terms in order to penalize deviations from the desired final water level of largest reservoir in the system (Miyun reservoir), from the desired final average hydraulic head of the groundwater storage as well as the deficit of the customer demand separated into two groups for household/industry and agriculture. The deficit of the delivered water must be less than 5 % for domestic/industrial clients and less than 25 % for agricultural clients. For the third scenario it is assumed that water can be transferred to the considered region up to an annual amount of 300 Mio. m^3 starting from the third year within the optimization horizon.

The overall water demand exceeds noticeable the natural sources. The reduced groundwater model has been derived under the assumption that this over-consumption was covered by the groundwater storage. This leads to a strong reduction of the average hydraulic head of the groundwater storage of about 9 meters during 5 years (scenario 1, see Fig. 10). Using the initial state of the Miyun reservoir and the final value of the reference trajectory for the average groundwater head as target in the objective function, for the base scenario (scenario 1) there are only small deviations from these values in combination with a minor demand deficit observable. In the second scenario the increase of the target value for the final groundwater hydraulic head at 5 m and a corresponding shift of penalty coefficients in order to keep this value lead to a better spreading of the overdraft over the different storages as well as to the customers.

Fig. 9 shows the course of the water level for two reservoirs. Because of its capacity of 4.4 10^9 m^3 the Miyun reservoir plays an important role for the long term management of the overall system. As can be seen, the years with above-average precipitation produce only a medium rise of the water level. The second scenario, which attempts to reduce the decline of the average hydraulic head of the groundwater storage, results in an increased release of this reservoir compared to the base scenario, which corresponds to a change of the final water level from 142 m above sea level to 137 m. In the third scenario a part of this release is replaced by water from outside of the considered region, which reduces the water level decrease at about 2 m. The second reservoir is situated at a channel from the Miyun-reservoir to the city of Beijing and serves only as intermediate storage. The admissible range for water management is completely utilized by the optimal control approach. The shift in the management target causes a different operating strategy because water from the connected channel is taken to replace groundwater abstractions in the nearby regions. The change of the management target for scenario 2 and scenario 3 induce also an increase of the overall demand deficit from 0.1 % to up to 3.5 % (scenario 2), which is nearly 6.8 10^8 m^3 over the full horizon.

Fig. 10 shows the time plot of the mean groundwater level of the optimization scenarios. It is obvious that in scenarios 2 and 3 the aimed increase of the target value for the final groundwater hydraulic head at 5 m is nearly achieved. In Fig. 11 the impact of the different strategies to the exemplary input 1 (exploitation in a certain region) and exemplary output 5 (groundwater level at a defined observation point) can be studied. The exploitation is clearly decreased in scenarios 2 and 3 which correspond to an increase of the groundwater level at the observation point. Finally, from Fig. 12 it can be seen that the performance of the drastically reduced groundwater model is good, reflecting the fact that the original FEM model with more than 100,000 nodes has been reduced to a state space model with 36 states.

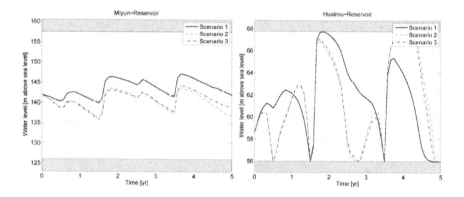

Figure 9. Water level of the largest reservoir in the water distribution system (Miyun-reservoir) as well as of a reservoir with seasonal storage capacity (Huairou-reservoir).

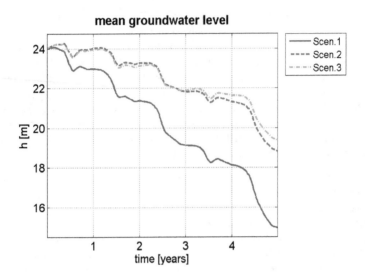

Figure 10. Time plot of the mean groundwater level in the optimization scenarios 1 – 3.

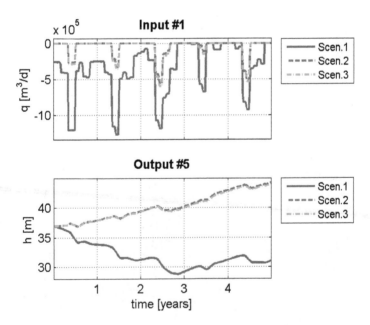

Figure 11. Time plot of input 1 (exploitation in a certain region) and output 5 (groundwater level at a defined point) in the optimization scenarios 1 – 3.

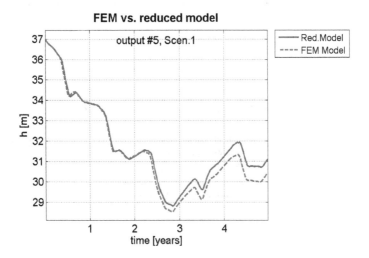

Figure 12. Time plot of output 5 of the full FEM model compared with the reduced model (optimization scenario 1).

6. Conclusions

In this paper an optimal control approach as a component of a decision support system (DSS) for the management of the total water resources (surface water, groundwater, external water resources) in a fast developing region under a critical water shortage has been presented. It has been proven that in spite of the large area which has to be managed and the corresponding complex surface and groundwater models the optimization problem could be solved in an appropriate computation time (~ minutes). This could be achieved by a drastic reduction of the complex groundwater model to a state space model of relatively low dimension (n < 50). The user (i.e. water allocation decision maker) is enabled to select from a number of predefined performance criteria as well as to assign constraints to the elements of the water allocation system in order to specify the management targets according to his/her needs. The performance of the proposed concept is demonstrated by close to reality optimization scenarios, whereby the benefit of a new strategic channel has been investigated with a planning horizon of 5 years. Actually the developed DSS component is used in a first version by the decision makers. Future work will focus on the application and adaptation of the developed concept and software for the water resources management of further regions with critical water shortage.

Acknowledgements

This work was supported by the German Federal Ministry of Education and Research (BMBF).

Author details

Thomas Bernard[1], Oliver Krol[1], Thomas Rauschenbach[2] and Divas Karimanzira[2]

1 Business Department MRD, Fraunhofer Institute of Optronics, System Technologies and Image Exploitation IOSB, Karlsruhe, Germany

2 Fraunhofer Application Center System Technology AST, Fraunhofer Institute of Optronics, System Technologies and Image Exploitation IOSB, Ilmenau, Germany

References

[1] Bernard T., Krol O., Linke H., Rauschenbach T. Optimal Management of Regional Water Supply Systems Using a Reduced Finite-Element. at- Automatisierungstechnik, 2009; (57)12: 593-600.

[2] Rauschenbach T., Steusloff H., Birkle M. Modeling of Beijing municipality's water resources as basis for an optimal water allocation system. In: Proceedings of Beijing Conference on Sustainable Water Management. Beijing, China; 2007.

[3] Mays L. W. Water resources system management tools. New York: McGraw-Hill Companies; 2005.

[4] McKinney D. C. International Survey of Decision Support Systems for Integrated Water Management. Technical Report, IRG Project No. 1673-000, Bucharest, 2004.

[5] Rauschenbach T.WaterLib 2.0 - Simulation and Management of Surface WaterSystems. In: 2008 Promotion Conference on International Advanced Water Technology.Beijing, China, 2008.

[6] Bear J. Dynamic of Fluids in Porous Media, New York, Dover Publications; 1988.

[7] Software package FEFLOW: http://www.feflow.info/ (accessed 5 June 2012)

[8] Bernard T., Linke H., Krol O. A Concept for the long term Optimization of regional Water Supply Systems using a reduced Finite Element Groundwater Model, VDI-Berichte 1980, Düsseldorf: VDI-Verlag, 2007; p751-762

[9]   ChenJ.Y., TangC.Y., ShenY.J., SakuraY., KondonA.,ShimadaJ. Use of water balance
      calculationand tritium to examine the dropdown of groundwatertable in the pied-
      mont of the north china plain (ncp). Environmental Geology, 44:564-571, 2003.

[10] Z. Han. Groundwater resources protection andaquifer recovery in china. Environ-
      mental Geology,44:106-111, 2003.

[11] Zippel M., Hannappel S. Ermittlung des Grundwasserdargebotes der Berliner Was-
      serwerke mittels regionaler numerischer Grundwasserströmungsmodelle. Grund-
      wasser, 4:195-207, 2008.

[12] FosterS., GardunoH., EvansR., OlsonD., TianY.,ZhangW., HanZ. Quarternary aquifer
      of thenorth china plain. Hydrogeology Journal, 12:81-93,2004.

[13] Antoulas A. C. Approximation of Large-Scale Dynamical Systems. Philadelphia:
      SIAM Press, 2005.

[14] Wolfrum P., Vargas A., Gallivan M., Allgöwer F.: Complexity reduction of a thin film
      deposition model using a trajectory based nonlinear model reduction technique.
      American Control Conference, Portland, USA, 2005.

[15] Wächter A., Biegler L. T. On the implementation of a primal-dual interior point filter
      line search algorithm for large-scale nonlinear programming. Mathematical Pro-
      gramming, 2006: 106(1):25–57.

[16] Franke R., Arnold E. The solver Omuses/HQP for structured large-scale constrained
      optimization: algorithm, implementation and example application, Sixth SIAM Con-
      ference on Optimization, Atlanta, USA, 1999.

[17] Griewank A., Juedes D., Utke J. ADOL-C: A package for the automatic differentiation
      of algorithms written in C/C++. ACM Trans. Math. Software, 1996: 22(2): 131-167.

Chapter 4

Energy Efficiency in Water Supply Systems: GA for Pump Schedule Optimization and ANN for Hybrid Energy Prediction

H. M. Ramos, L. H. M. Costa and F. V. Gonçalves

Additional information is available at the end of the chapter

1. Introduction

In the last decades, the managers of water distribution systems have been concerned with the reduction of energy consumption and the strong influence of climate changes on water patterns. The subsequent increase in oil prices has increased the search for alternatives to generate energy using renewable sources and creating hybrid energy solutions, in particular associated to the water consumption.

According to Watergy (2009), about two or three percent of the energy consumption in the world is used for pumping and water treatment for urban and industrial purposes. The consumption of energy, in most of water systems all over the world, could be reduced at least 25%, through performance improvements in the energy efficiency. Hence, it is noticeable the importance of development of models which define operational strategies in pumping stations, aiming at their best energy efficiency solution.

The consumption of electric energy, due to the water pumping, represents the biggest part of the energy expenses in the water industry sector. Among several practical solutions, which can enable the reduction of energy consumption, the change in the pumping operational procedures shows to be very effective, since it does not need any additional investment but it is able to induce a significant energy cost reduction in a short term. As well known, the tasks of operators from the drinking network systems are very complex because several distinct goals are involved in this process. To determine, among an extensive set of possibilities, the best operational rules that watch out for the quality of the public service and also provide energy savings, through the utilization of optimization model tools which take into consideration all the system parameters and components, is undoubtedly a priority.

The technological advances in the computational area enabled, in the last years, the intensification of the quality of scientific works related to the optimization tools, as well as aiming at the reduction of the energy costs in the operation of drinking systems. Nevertheless, most of the optimization models developed was applied to specific cases.

The first studies to optimize the energy costs of pumping have been used for operational research techniques, such as linear programming (Jowitt and Germanopoulos, 1992), integer linear programming (Little and Mccrodden, 1989), non-linear programming (Ormsbee et al., 1989) and dynamic programming (Lansey and Awumah, 1994). The limitation of using these models to real cases is mainly due to the complexity of the equations' resolution to ensure the hydraulic balance and the difficulty of generalizing such optimization models in any water supply system (WSS).

Brion and Mays (1991), in the attempt to reduce the operational costs in a drinking pipeline in Austin, Texas (USA), had tested a model of optimization and simulation, achieving a reduction of 17.3 % in the operational costs. Ormsbee and Reddy (1995) applied an optimization algorithm in Washington - DC and obtained significant results with the management implementation provided by the model, observing a reduction of 6.9% in the costs with electric energy. During this period, the use of evolutionary algorithms was quite limited. Wood and Reddy (1994) were the pioneers in the use of such algorithms.

The remarkable use of evolutionary algorithms in this research topic in recent years is mainly due to Genetic Algorithm (GA) provides a great flexibility in exploring the search space and allows an easy link to other simulation models. However, in contrast the GA does not solve problems with constraints. Once the operation in WSS is considered a complex procedure, with many constraints, there remains the doubt about the speed of the modelling and the convergence for optimal solutions between the GA and hydraulic simulators.

Additionally the concern with the reduction of the computational time is due to the applicability of energy optimization models in real time (Martinez, et al., 2007; Jamieson, et al., 2007, Salomons et al., 2007; Rao and Alvarruiz, 2007; Rao and Salomons, 2007; Alvis et al., 2007). To reduce the computational time for seeking solutions with reduced energy costs, these authors used the technique of Artificial Neural Networks (ANN) to reproduce the results by the hydraulic simulator obtained by the EPANET (Rossman, 2000). Then, this new tool based on ANN for the hydraulic simulation was connected with a GA model. After several analyses done in a hypothetical system and in two real case studies, the authors concluded the model GA-ANN found optimal solutions in a period 20 times lower when compared to GA-EPANET. Shamir and Salomons (2008) have searched for reducing the computational simulation time based on a scale model of a real case system for different operating conditions.

At the present research a different resolution was adopted. In order to reduce the computational simulation time in the search for optimal solutions, a change in the GA algorithm type was made, instead of replacing the hydraulic simulator model (EPANET) as former references. Thus, new algorithms were created which work directly with the infeasible solutions generated by a GA to make them feasible, through the development of a hybrid genetic algorithm (HGA) (i.e. genetic algorithm plus repair algorithms).

This new model determines, in discrete intervals (every hours) the best programming to be followed by the pumps switch on / off, in a daily perspective of operation. In this way, the decisions start to be orientated from the research of thousands of possible combinations, being chosen, through an iterative process, the best energy management strategy that presents the best energy savings.

The world's economy is directly connected to energy and it is the straight way to produce life quality for society. China is nowadays one of the biggest consumers of energy in the world (Wu, 2009). In order to have enough energy to make its economy grow the prediction of new solutions to produce sustainable energy in a most feasible way is imperative, not only depending on conventional sources (i.e. fossil fuel) but using renewable sources. The increase of energy consumption and the desired reduction of the use of fossil fuels and the raise of the harmful effects of pollution produced by non-renewable sources is one of the most important reasons for conducting research in renewable and sustainable solutions. In Koroneos (2003) analysis, renewable sources are used to produce energy with high efficiencies, social and environmental significant benefits.

Renewable energy includes hydro, wind, solar and many others resources. To avoid problems caused by weather and environment uncertainties that hinder the reliability of a continuous production of energy from renewable sources, when only one source production system model is considered, the possibility of integrating various sources, creating hybrid energy solutions, can greatly reduce the intermittences and uncertainties of energy production bringing a new perspective for the future. These hybrid solutions are feasible applications for water distribution systems that need to decrease their costs with the electrical component. These solutions, when installed in water systems, take the advantage of power production based on its own available flow energy, as well as on local available renewable sources, saving on the purchase of energy produced by fossil sources and contributing for the reduction of the greenhouse effect. In recent studies (Moura and Almeida, 2009; Ramos and Ramos, 2009a; Ramos and Ramos, 2009b; Vieira and Ramos, 2008, 2009), the option to mix complementary energy sources like hydropower, wind or solar seems to be a solution to mitigate the energy intermittency when comparing with only one source. So, the idea of a hybrid solution has the advantage of compensating the fluctuations between available sources with decentralized renewable generation technologies.

In literature review, a sustainable energy system has been commonly defined in terms of its energy efficiency, its reliability, and its environmental impacts. The basic requirements for an efficient energy system is its ability to generate enough power for world needs at an affordable price, clean supply, in safe and reliable conditions. On the other hand, the typical characteristics of a sustainable energy system can be derived from policy definitions and objectives since they are quite similar in industrialized countries. The improvement of the efficiency in the energy production and the guarantee of reliable energy supply seem to be nowadays common interests of the developed and developing countries (Alanne and Saari, 2006).

This work aims to present an artificial neural network model by the optimization of the best economical hybrid solution configuration applied to a typical water distribution system.

2. Models formulation

2.1. Objective function

The search for the optimal control settings of pumps in a real drinking network system is seen as a problem of high complexity, due to the fact that it involves a high number of decision variables and several constraints, particular to each system. The decision variables are the operational states of the pumps xt (x 1t, x 2t, ..., x Nt), where N represents the number of pumps and t is the time-step throughout the operational time.

To represent the states of the decision variables in each time-step, the binary notation was used. The configuration of each pump is represented by a bit where 0 and 1 stated switched on and off, respectively. The main goal of the model is to find the configuration of the pumps' status which proceeds to the lowest energy cost scenario for the operational time duration. To calculate this cost, several variables must be considered, in each time-step, such as the variation of consumption, energy tariff pattern and the operational status of each pump.

The objective function is the sum of energy consumed by the pumps, in every operational time, due to the water consumption and tanks' storage capacity. It can be expressed according to the following equation:

$$Minimize \sum_{n=1}^{N} \sum_{t=1}^{24} C_{nt} E_{nt}(X_{nt})[1]$$

(1)

where E and C stated the consumed energy (kWh) and the energy costs by pumps' operation in the time-step t.

2.2. Constraints

The main constraint of the model is the hydraulic balance verification for the network. To establish such balance, the equations of the conservation of mass at each junction node and the conservation of energy around each loop in the network are satisfied. In order to these conditions be attended it is necessary to accomplish the hydraulic verifications to each system configuration. The hydraulic simulator EPANET (ROSSMAN, 2000) was used to perform this purpose.

The constraints are implicit in the calculation of the objective function. These are equations that need to be solved in order to obtain the total energy cost of the solution to be analyzed. After accomplishing this stage, some variables are verified, from the hydraulic simulation, aiming for obtaining the hydraulic performance of the system that it is evaluated by means of explicit constraints, showed as follows:

Pressure: for each time-step of operational time, the pressures in all the junction nodes must be between the minimum and maximum limits.

$$P\min_i \le P_{it} \le P\max_i \qquad \forall_i, \forall_t \tag{2}$$

where P_{it} represents the pressure on node i in time-step t, Pmin $_i$ and Pmax $_i$ are the minimum and maximum pressures required for node i.

Levels of storage tanks: The levels of storage tanks must be between the minimum and maximum limits for each time-step. Besides at the end of the operational time duration, they must be superior to the levels at the beginning of the time duration. This last constraint assures the levels of the tanks do not lessen with the repetitions of the operational cycles.

$$S\min_j \le S_{jt} \le S\max_j \forall_j, \forall_t \tag{3}$$

where S_{jt}: level of tank j in time-step t; Smin $_j$ e Smax $_j$: minimum and maximum levels of storage tank j.

$$S_j(24h) \le S_j(0h) \forall_j \tag{4}$$

Pumping power capacity: the power used by each pump during the operational time must be inferior to its maximum capacity.

$$PP_{kt} \le PP\max_k \forall_k \tag{5}$$

where PP_{kt}: used power by pump k in time-step t; PPmax$_k$: maximum capacity of the pump k.

Actuation of the pumps: The number of pumping start-ups in the operational strategy must be inferior to a pre-established limit. This constraint, presented by Lansey and Awumah (1994), influences in the maintenance of each pump, since the more it is put into action in a same operational cycle, the bigger will be its wear. Lansey and Awumah (1994) suggest the maximum pump start-ups 3 in 24 hours. A greater value can cause problems on the pumps inducing the need of maintenance and repair and consequently the interruption of the system operation.

$$NA_k \le NA\max_k \tag{6}$$

where: NA_k represents the number of start-ups for pump k and NAmax$_k$ the maximum allowable pump start-ups for the pump k.

2.3. Optimization algorithm

The definition of optimal control strategies in water distribution systems, where the rules evaluate the behaviour of the system and make decisions at each time-step, requiring a great computational demand. Among several available optimization methods, the Genetic Algorithm (GA) was the tool chosen for offering a great flexibility in search space, allied to the possibility of use discrete variables. Besides these advantages, the technique has an easy manipulation, which makes its connectivity with simulation models easier.

The model developed is composed by two modules that will work as a whole in a way the hydraulic simulation routine is called to simulate each operational alternative scenarios given by the GA, in the search of alternatives with better performance.

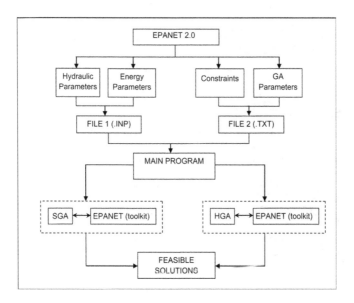

Figure 1. Stages of the optimization model

Further a Simple Genetic Algorithm (SGA), a Hybrid Genetic Algorithm (HGA) was also developed. This algorithm was built from a combination of a conventional GA with a method of correction of solutions and a specialized local search procedure. The goal is to find, in a faster way, feasible solutions, which are difficult to be found by traditional genetic algorithms due to the tendency that the situation has to generate a high number of impracticable hydraulic solutions.

The flowchart containing the steps of the optimization model is shown in the Figure 1.

2.4. Prediction algorithm

The conception of an ANN in order to capture the best energy model domain from a configuration model and economical simulator (CES) in a much more efficient way is based on the
following remarks: first of all, a robust data base has to be developed to create the input and
output data set that will be used in ANN conception and training; the data has to be analysed to determine a structure that fits the problem and then to train and validate the ANN.

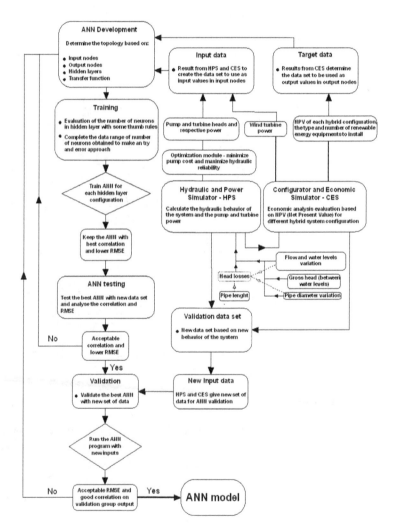

Figure 2. Flowchart for the developed ANN model.

A flowchart describing the procedures of the designed ANN is shown on Figure 2.

The data used on this study is calculated by means of a CES model that gives an optimized ranking of the best hybrid solution for each particular case, based on an economy analyses for the production and consumption of energy (Figure 2). This data set is organized with the subject that the study is concerned to evaluate the use of hybrid energy solutions in water distribution systems based on micro-hydro, wind turbine and national electric grid. Hence, the range of data is defined in order to adequate the installation of such energy converters. The data range for flow, power head and water levels variation in reservoirs are used in a hydraulic and power simulator (HPS) to determine the power consumed by the pump and the power produced in a micro-hydro turbine installed in a gravity pipe branch whenever there is energy available in the system.

3. Methodology

3.1. Simple Genetic Algorithm (SGA)

GA is a stochastic method of global search that develop such search through the evolution of a population, where each element (or individual) is the representation of a possible solution for the problem. The principle is based on the theory of natural selection and it was firstly presented by Goldberg (1989).

At drinking systems' operation, GA stands out for being very efficient when binary and discrete variables are used. They represent a set of optimal solutions and not only one. At each new computational step, solutions containing the status of the pumps are evaluated and later classified according to its fitness. The tendency is as the running proceeds, the elements with less fitness disappear and the more adapted to the impositions (or constraints) of the problem will arise.

GAs do not deal directly with the optimization problems that contain constraints. This impediment in the minimization procedure can be overcome employing the Penalty Methods, on which pre-defined constraints are added to the objective function in terms of penalties, turning the solution less apt as much as its violations occur. The Multiplicative Penalty Method (MPM), presented by Hilton & Culver (2000), is then implemented in this model. The penalty function is presented as follows:

$$P_{TR} = \prod_{i=1}^{NTR} k \qquad (7)$$

where TR: type of constraint; NTR: amount of hydraulic elements (nodes, reservoirs or pumps) which have violated certain constraints; k: coefficient which varies with the hydraulic element and the type of violated constraint.

Table 1 shows the values of k depending on the type of violated constraint.

TR	Hydraulic Element	Violated Constraint	k
N1	Nodes	Pressure between the limits (min. and max.)	1.05
N2	Nodes	Positive pressure (continuity of supply)	1.80
R1	Tanks	Water level between the limits (min. and max.)	1.20
R2	Tanks	Water level at 24h greater than the initial level	1.50
B1	Pumps	Maximum capacity of pump	1.20
B2	Pumps	Number of actuations	1.50

Table 1. Values of k

The values of k represent how the energy cost is increased for a particular type of violated restriction (TR). These values were determined from the amount and importance of constraints in the model. Analyzing the extreme values (1.05 and 1.80), for each node that exceed their limits, increases 5% to the value of the objective function. It was adopted the lower value for this violation because, commonly, the number of nodes in a WSS is higher the amount of tanks and pumps. However, as the discontinuity of the supply occurs in the system, it has great importance in the feasibility of the solution consequently a maximum value was adopted for this type of violation, increasing by 80% the cost of energy. Following this logic, the remaining violations have intermediate k values. When the constraint is not violated the coefficient k has the unit value.

The first stage of the SGA (Figure 1) process is characterized by the generation of operational rules (randomly), the demand definition and the tariff costs. Next, these variables are used by the hydraulic simulator (i.e. EPANET), which calculates the pressures in the pipe system nodes, the energy consumed and the levels of the tanks, all of them being necessary for the evaluation of the solution. The following stage is characterized by the calculation of the objective function, which is obtained from the total energy cost and from the penalty function, in case of infeasible solution. The process is repeated until the parameters of the operational control meets the hydraulic requirements with the lowest cost possible.

3.2. Hybrid Genetic Algorithm

SGA makes use of the penalty method becoming the infeasible solutions into solutions with reduced ability. The genetic operators only diversify the solutions, but do not become them feasible. In this case, it can be confirmed the search process for solutions hydraulically feasible, with minimum energy costs, is strongly stochastic. During the process of evaluation of the objective function, the explicit restrictive variables can be evaluated every hour. Thus, at this time interval, it is possible to verify the type of constraints that were violated. Because of this, repair algorithms were created, and every hour they try to correct the solutions generated by GA, becoming them hydraulically feasible. The HGA layout of the model is also presented at Figure 3. Hence, each solution generated by GA is passed on to the repair algorithms. After this stage two solutions are stored: the original, generated by GA, and the modified solution, generated after the attempts of correction. If the penalty function of the modified solution is zero, so it will be sent to a data bank, otherwise, this solution will be discarded. Independent on the destiny of the modified solution, the original solution will be conserved and sent to the next generations of the GA, avoiding a premature convergence of the solutions.

The repair algorithms are only a set of rules that modify the decision variables trying to become solutions hydraulically feasible all hours (Figure 4).

Among the type of corrections presented in Figure 4, the one related to the maximum number of pump start-ups is the only one that does not use the EPANET routines. This is the first type of repair that occurs in infeasible solutions and aims mainly the reduction of the pump start-ups, changing as little as possible the original configuration of the solution.

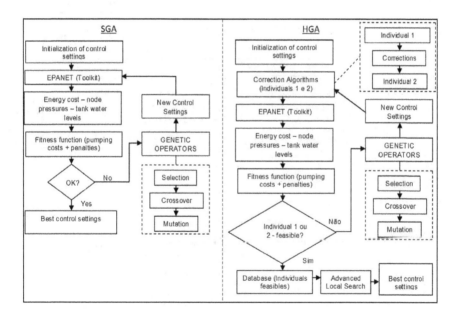

Figure 3. Flowcharts: SGA and HGA

Figure 5 illustrates this type of repair to a solution of a pump with six start-ups.

In Figure 5, with only four changes, it was reduced from six to two the number of start-ups. Besides the considerable reduction, in the repaired solution is visible a greater uniformity of pumps' switch-on schedules. The changed solution has presented only two periods with the pump switched-on. The use of long operation periods is a characteristic of commonly strategies in real pump systems due to a lesser intervention in the operation and a wear reduction of the pumps.

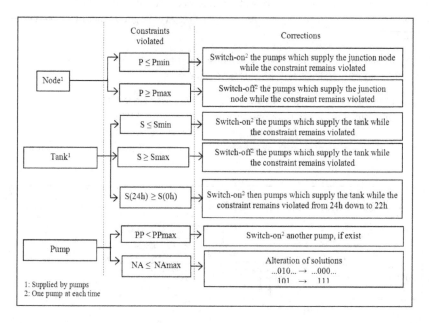

Figure 4. Type of corrections

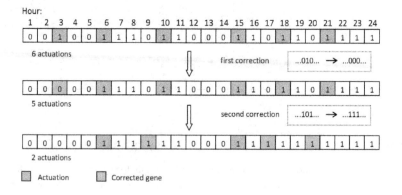

Figure 5. Example of correction – Actuation as start-up of the pumps

Finishing the iterations of the HGA, the solutions stored at the data bank (feasible solutions) are sent to a process of specialized local search. This search algorithm is an iterative process in which, every hour, the pumps are switched-off one by one, verifying if the constraints remain inviolate. If the solution becomes hydraulically unfeasible, the initial solution is restored. The selected hour is the one that has the highest energy cost. The process is repeated until there are no alterations that result in feasible solutions.

With the utilization of the specialized local search algorithm it is possible to evolve good solutions in local optimal solutions. These solutions would probably require great computational efforts to be found by the conventional GA.

3.3. Artificial Neural Network

The data of renewable sources performance characteristics is included in the CES model to determine the best hybrid energy solution to be selected. One of the resources data is the wind turbine power curve of a selected wind turbine, which corresponds to the local wind source along an average year for the region under analysis (Figure 6) and the wind annual average speed applied to the wind turbine. In Table 2 is presented an example of data set range to be used in the CES model to determine the inputs and outputs of the developed ANN. Those data is used to calculate all energy and economic parameters to be included in the CES model to complete the data needed to train the ANN.

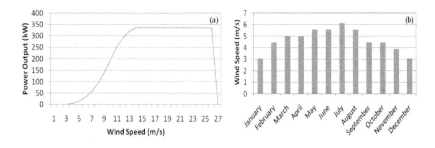

Figure 6. Wind energy: Wind Turbine Power Curve for an Enercon E33 and Wind source for one year at Lisbon region

Based on a basic data range, depending on the system characteristics (Table 2), to be used in the CES model and from auxiliary hydraulic and energy formulations, the complete input data is then obtained (Table 3) being: (1) Pump power (kW); (2) Pump energy consumption (kWh); (3) Turbine power (kW) - average output; (4) Flow (m3/s) - annual average flow; (5) Gross head (m); (6) Pumping head (m); (7) Head losses (m); (8) Power net head (m); (9) Design pumping flow rate (l/s); (10) Wind speed (m/s) - annual average; and (11) Wind turbine power (kW) - annual average output.

In the end of the modelling process the input data set is built in a matrix of [11 x 19,602] (Table 3), which by the interaction of the wind velocity data and the water flow yields in the output matrix of [5 x 19,602] (Table 4), representing the Net Present Value (NPV) of each hybrid solution configuration, as well as the number of wind turbines to be installed.

Wind speed annual average (m/s)	Flow (l/s)	Power net head (m)	Gross Head (m)
1.5	10	2	10
2.0	20	7	16
2.5	30	13	21
3.0	40	18	27
3.5	50	24	32
4.0	60	29	38
4.5	70	35	43
5.0	80	40	49
5.5	90	46	54
6.0	100	51	60
6.5	150	57	66
7.0	200	62	71
7.5	250	68	77
8.0	300	73	82
8.5	350	79	88
9.0	400	84	93
9.5	450	90	99
10.0	500	95	104
10.5	550	101	110
11.0	600	106	116
11.5	650	112	121
12.0	700	117	127
12.5	750	123	132
13.0	800	128	138
13.5	850	134	143
14.0	900	139	149
14.5	950	145	154
15.0	1000	150	160

Table 2. Basic data set range used in CES.

The ANN data set created to be used in water distribution systems is then ready to determine the NPV of each hybrid system evaluated for each type of configuration (e.g. grid, grid + hydro, grid + wind, grid + hydro + wind).

Pump power kW/h (1)	Pump primary load kW/d (2)	Turbine mean output power kW (3)	Annual average flow m³/s (4)	* Z m (5)	Pumping head m (6)	Head loss m (7)	Power head m (8)	Design flow rate L/s (9)	Wind speed m/s (10)	Wind turbine mean output power kW (11)
....	....	....	....	....	....	....	....	....	....	....
0.322	2.895	0.587	0.01	16	24	8	7	16	3	15
0.398	3.584	1.016	0.01	21	29	8	13	16	3	15
0.475	4.274	1.446	0.01	27	35	8	18	16	3	15
0.552	4.964	1.876	0.01	32	41	8	24	16	3	15
0.628	5.653	2.306	0.01	38	46	8	29	16	3	15
0.705	6.343	2.735	0.01	43	52	8	35	16	3	15
0.781	7.032	3.165	0.01	49	57	9	40	16	3	15
0.858	7.722	3.595	0.01	54	63	9	46	16	3	15
0.935	8.412	4.025	0.01	60	69	9	51	16	3	15
1.011	9.101	4.454	0.01	66	74	9	57	16	3	15
1.088	9.791	4.884	0.01	71	80	9	62	16	3	15
1.165	10.481	5.314	0.01	77	86	9	68	16	3	15
1.241	11.170	5.744	0.01	82	91	9	73	16	3	15
1.318	11.860	6.173	0.01	88	97	9	79	16	3	15
1.394	12.549	6.603	0.01	93	102	9	84	16	3	15
....	....	....	....	....	....	....	....	....	....	....

Table 3. Input data set for the system characteristics used in ANN.

Matlab® is used for the ANN development. The creation of an ANN should comprise the following steps: (i) patterns definition; (ii) network implementation; (iii) identification of the learning parameters; (iv) training, testing and validation processes. A new neural network model of hybrid energy must be compared with an energy configuration model and economical simulator (CES) using the following procedures: CES is used to obtain data applied in the training process and in reliable neural network tests, together with an hydraulic and power simulator model (HPS) for a large range of flow rates, gross heads, pumping and power heads and wind velocities. That data, available on Ramos and Ramos (2009b) research, uses the HPS to hydraulically balance the water distribution system, in a village of Portugal, determining the hydraulic behaviour of the all system including the most suitable pump and turbine operation for each flow condition.

NPV€ Grid	NPV€ Grid+Hydro	NPV€ Grid+Wind	NPV€ Grid+Hydro+Wind	Wind Turbine Installed
-59.00	1812.00	-571464.00	-569553.00	1
-78.00	6617.00	-571495.00	-564747.00	1
-96.00	12391.00	-571526.00	-558973.00	1
-115.00	17197.00	-571557.00	-554168.00	1
-133.00	22971.00	-571588.00	-548394.00	1
-152.00	27776.00	-571619.00	-543588.00	1
-170.00	33550.00	-571650.00	-537814.00	1
-189.00	38356.00	-571680.00	-533009.00	1
-207.00	44130.00	-571712.00	-527235.00	1
...	...	...	...	...
-226.00	48935.00	-316043.00	-266690.00	2
-244.00	54710.00	-316077.00	-260916.00	2
-263.00	59514.00	-316111.00	-256110.00	2
-282.00	65289.00	-316146.00	-250337.00	2
-300.00	70094.00	-316180.00	-245531.00	2
-319.00	75868.00	-316214.00	-239757.00	2
-337.00	80674.00	-316248.00	-234951.00	2
-356.00	86447.00	-316282.00	-229177.00	2
-374.00	91253.00	-316317.00	-224372.00	2
...	...	...	...	...
-393.00	97027.00	109886.00	207679.00	3
-411.00	101832.00	109850.00	212483.00	3
-430.00	107606.00	109813.00	218258.00	3
-448.00	112411.00	109778.00	223062.00	3
-467.00	118185.00	109741.00	228838.00	3
-485.00	122991.00	109706.00	233644.00	3
-504.00	128765.00	109669.00	239416.00	3
-522.00	133570.00	109633.00	244223.00	3

-541.00	139344.00	109597.00	249997.00	3
-559.00	144150.00	109561.00	254802.00	3

Table 4. Input data set for the best economic configuration used in ANN.

In the ANN code running, the process of training and simulation for each system characteristic is analysed. In the training mode is introduced the configuration parameters. Those parameters are standard limits (max and min), number of neurons on the hidden layer, limit number of epochs, final error desired, validation rate and activation function used in the hidden layer. With the best ANN configuration for each possible hybrid system and new data set for inputs, a validation process is made and the results are verified in terms of correlation and relative error among the values of CES base model and the ANN.

4. Case studies

4.1. Optimization of the pumps' schedule in the Fátima system

The drinking system of Fátima is composed of 22 water sources, 10 treatment plants, 36 pump-stations and 64 tanks. The water is distributed to the consumers through 1111 km by a supply and distribution network system. Nowadays, the system is managed by the company Veolia – Águas de Ourém, which is responsible for the catchment, water treatment and distribution (Figure 7).

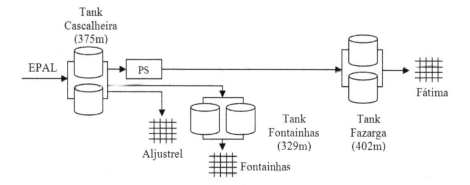

Figure 7. Drinking system of Cascalheira's tank

The supply system chosen for this case study supplies the tank Fazarga with an elevation of 402 m. This tank is responsible for the service to the demands of the region of Fátima and other close locations. This supply system has a pump station (PS) located in the proximities of the tank Cascalheira (elevation: 375 m). This last one is supplied by EPAL (Portuguese Lisbon Water Company) and provides water, by gravity, to the locations of Aljustrel and Fontainhas.

According to former description, the water storage of the tank Cascalheira is done by EPAL. The cost attributed to Veolia by this supply is related only to the effluent volume from this tank and it is not dependent of any alteration in the operation of the pump-station between tanks of Cascalheira and Fazarga. The reduction of this cost would only be possible with the implementation of water loss control by leakage. The level of the tank Cascalheira is always maintained close to the maximum limit in a way that it increases the reliability of the system. Thus, in the optimization model, it was chosen to consider only the variation of the level of the tank Fazarga at downstream of the pump-station.

The tank of Cascalheira has the storage capacity of 4000 m^3 of water, whereas Fazarga has a total volume of 347 m^3 and operates with the initial, minimum and maximum levels of 2.0 m, 0.3 m and 2.3 m, respectively. The pump-station comprises two pumps of Grundfos NK65-250 type which work for an average flow of 42 1/s with an efficiency of 65%.

The average time variation of the consumption in the region of Fátima during the day was obtained from the sensors located at the exit of the tank Fazarga. The period analyzed was from March to September, 2007. The water consumption in this year is more noticeable for comprising spring and summer. Figure 8 presents the average time variation calculated.

The hours with the pump working are considered as regular and discrete intervals by the optimization algorithm. Thus, for this case study, a day in which the pumps remained switched-on, in intervals similar to the format considered in the optimization model, were chosen. The hydraulic model of the system was built, in which the tanks Cascalheira and Fazarga were considered as reservoir and storage tank, respectively.

The variation in the level of the tank of Fazarga during the day calculated by the hydraulic simulator was similar to the real values. The maximum number of pump start-ups (Na max) used by Veolia was three (pump 1) and the level of the tank at the end of the operational time is very close to the initial one (Figure 9). The variation of the energy rate is presented in Table 5.

Hour	1:00	2:00	3:00	4:00	5:00	6:00	7:00	8:00	9:00	10:00	11:00	12:00
Tariff	0,0465	0,0465	0,0465	0,0465	0,0465	0,0465	0,0465	0,0465	0,0465	0,0761	0,1299	0,1299
Hour	13:00	14:00	15:00	16:00	17:00	18:00	19:00	20:00	21:00	22:00	23:00	24:00
Tariff	0,1299	0,0761	0,0761	0,0761	0,0761	0,0761	0,0761	0,0761	0,1299	0,1299	0,0761	0,0465

Table 5. Hour vs Energy Tariff (€/kWh) for Fátima system

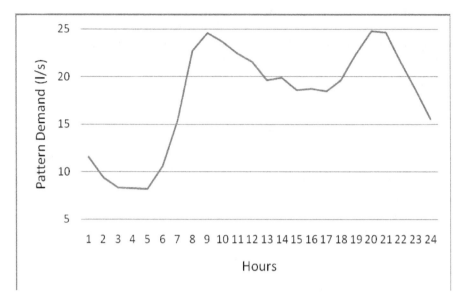

Figure 8. Pattern demand of Fátima system

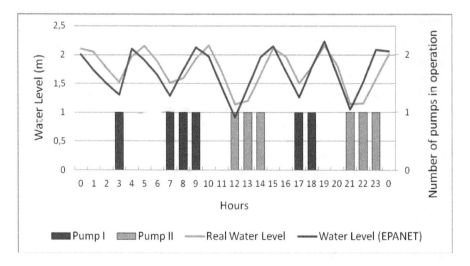

Figure 9. Control pump strategy

Both GA models presented in this analysis were implemented to determine the best operational strategy with a reduced energy cost in the system Cascalheira/Fazarga. Figure 10 presents the evolution of the objective function with the computational time, in minutes.

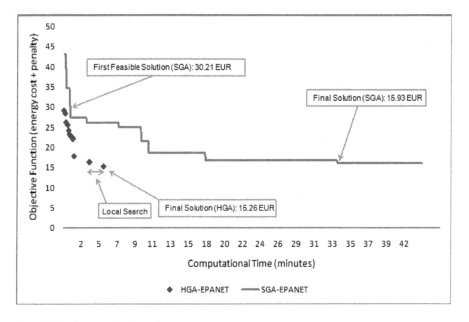

Figure 10. Convergence of the fitness functions

It is possible to evaluate the efficiency of the HGA model. Only with the feasible solutions obtained with 20 generations, from the repair algorithms and from the specialized local search system, it is possible to find a local optimal solution in about 5 minutes, whereas the SGA took a little more than 33 minutes to find a good solution, with also a bit higher energy cost when compared to the solution found by the HGA. The difficulty for GA to find a good feasible solution can quickly be confirmed. Such behaviour occurs due to the high level of randomness existent in GA models. The alterations of the solutions provided by the genetic operators diversify the type of answer without a guarantee of the evolution in each generation. Among all possible solutions, the probability of extracting, for each pump, a solution with at most three start-ups is 0.0173. Now, it is possible to confirm the difficulty of obtaining a feasible solution, because besides the determination of a solution it is necessary the other constraints (pressure limits, water levels in tanks and power pumps start-ups) be satisfied. These constraints are dependent on the complexity of the drinking system to be evaluated.

The energy cost due to the operation was 22.22 euros (date: 07 (day)/12 (month)/07 (year)). The pumps remained switched-on during 12 hours. A period of two hours (13h and 22h) belongs to the period with the most expensive energy tariff (Figure 11). The variation in the reservoir level is the main factor in the decision making the operation and the variation of the energy tariff is the second reason.

The best solution obtained by HGA, in each iteration step, is selected from a set of solutions containing only individuals hydraulically feasible. The objective function for this case is the total energy cost. For SGA while the model does not find a feasible solution, the objective function starts to be the sum between the energy cost and the penalty function. The operational strategy found by the HGA and the variations of the water level in the Fazarga tank for the real situation and the solution with reduced energy cost are shown at Figures 11 and 12. From Figures 9, 11 and 12 it is possible to make a comparison between the operational strategies presently adopted by the water manager company and the one obtained by HGA optimization model. The variation of the energy tariff was well explored in the solution with an important reduction of the energy cost (HGA). It is possible to observe a significant difference from the strategies, being noticeable that the pumps do not work in hours with energy tariff more expensive. With the implementation of the optimization model an economy of 31% was achieved for the period chosen for the analysis.

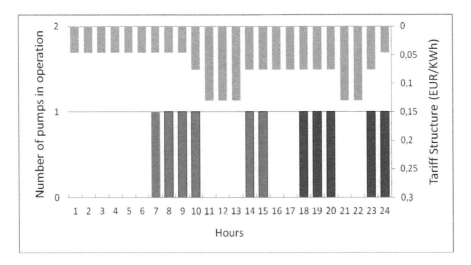

Figure 11. Control pump strategy (HGA).

In operational terms, the strategy obtained from the HGA can be considered more daring. In the critical time (1:00 p.m.) the level of the tank in the present operation by the water company achieved values superior to 1m. However based on former mentioned, the minimum water level in the Fazarga tank is 0.30m. In case of desirable an economic solution with higher levels in the tanks, it is easy to increase the minimum limit of the water level in the constraints of the HGA developed model. The importance between the minimum water level attained in the tank and the energy costs to be paid by the water company will depend on the water company priorities, economic and social impacts, and performance or feasibility factors.

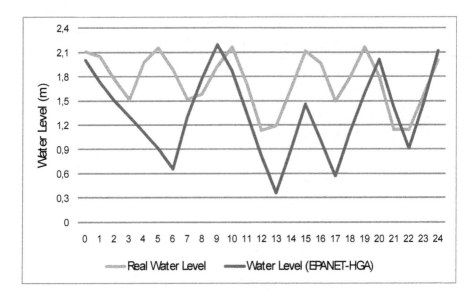

Figure 12. Water level of Fazarga tank

4.2. Prediction of hybrid energy solutions in Espite system

Espite is located in Ourém and it is a small system that distributes water to Couções and Arneiros do Carvalhal villages and the average flow in this pipe system is approximately 7 l/s. This system is hydraulically analysed to determine the best hydro solution. Then ANN is applied to establish the best economical hybrid solution, employing the same data set used to developed ANN model. A simplified scheme of Espite water drinking system is presented in Figure 13.

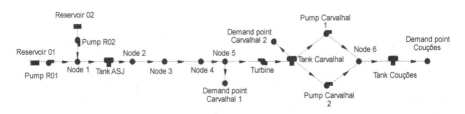

Figure 13. Scheme of Espite water distribution system

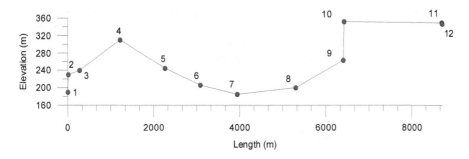

Figure 14. Elevation and length profile of Espite pipeline

The pump station considered in the analysis is Pump Carvalhal 1 and 2 and the micro hydro power plant will be installed in the gravity pipe system between node 5 and Tank Carvalhal. The population consumption (i.e. demand points) must be guaranteed and the tanks water level variation should vary between recommended limits. The elevation profile of Espite system is established in Figure 14, where (1) Reservoir 01; (2) Pump R01; (3) Node 1; (4) Tank ASJ; (5) Node 2; (6) Node 3; (7) Node 4; (8) Node 5; (9) Turbine, Tank Carvalhal, Pumps Carvalhal 1 & 2,; (10) Node 6; (11) Tank Couções and (12) Demand point Couções.

The HPS model is used to verify all hydraulic parameters and the system behaviour when a hydropower is installed. Rule-based controls are defined in the optimisation process to guarantee that the limit tank levels are always respected. In order to determine the most adequate hydro turbine in this water pipe system, regarding the importance to always maintain a good system operation management and the satisfactory demand flows, the evaluation of the available energy and the characteristic turbine curve compatible with the all operating and hydraulic constrains must be developed. According to Araujo (2005) and Ramos et al. (2010), a characteristic curve for the turbine is evaluated to define the most adequate turbine selection a key for the successful of this solution. The system is then analysed using the electricity tariff for the worst conditions. The energy report of the original situation is shown in Table 6.

	Energy Report		
Pump Station	Use*(%)	Consumption kWh/m³	Max. Power kW
Carvalhal 1	100,00	0,78	4,51
Carvalhal 2	100,00	0,78	4,51

Table 6. Pump cost with original situation. *basis reference

To reduce the pump consumption, the optimization of the time pumping is considered, turning it on in the low electricity tariff period and turning it off in the higher tariff period, always imposing tank levels' restriction to satisfy the minimum and maximum advisable values for its good operation. Figure 15 shows the system behaviour regarding the water level variation and the optimized pump operation time. Table 7 shows the savings achieved with the water level control and pump operation optimization for the energy tariff pattern adopted.

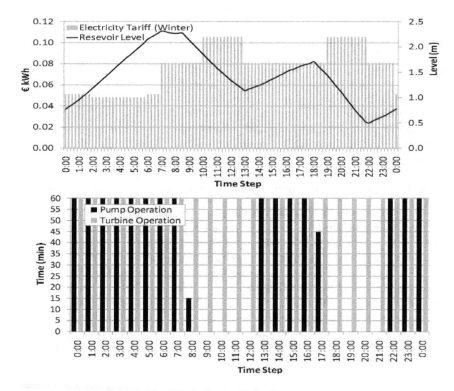

Figure 15. System behaviour with reservoir level control and pump operation optimization: water level variation in Couções tank, electricity tariff and pump and turbine operation time.

	Energy Report			
Pump Station	Use*(%)	Consumption kWh/m³	Max. Power kW	Saving (%)
Carvalhal 1	65.09	0,55	3.24	58.19
Carvalhal 2	65.09	0,55	3.24	58.19

Table 7. Pump benefits with optimization of water level control and pump operation

The energy production in the hydro power is calculated using the hydraulic turbine selected considering a sell rate of 0.10€/kWh for 24 hour production as shown in Table 8 as well as the saving achieved with this energy configuration. The operating point of the turbine corresponds to a power net head of 40 m and an average flow of 6.6 l/s determined by the HPS model based on extended period simulations of 24h.

	Energy Report			
Turbine	Production kWh/m³	Max. Power kW	Power/day kW	Saving (%)
Carvalhal	0.07	2.12	49.04	63.35

Table 8. Energy production in the hydropower solution.

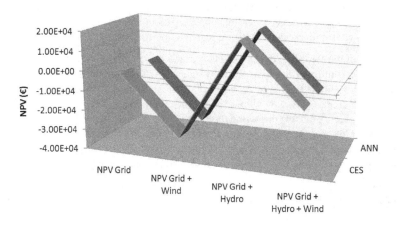

Figure 16. NPV results by ANN and CES models for the case study.

After the calculation of the pump consumption and the turbine production, the values are inserted in the ANN model developed and compared with the results obtained with the CES model. For the analysis of the best hybrid energy solution it takes into account that the wind speed in the region of this case study has an average value of 5 m/s. It was considered the wind turbine model SW Skystream 3.7 with a rated power of 1.8 kW and a market price of € 15,000 and a micro hydro turbine (or a pump as turbine – PAT) with a market price estimated in € 2,500 with a nominal power of 3.14 kW. For a lifetime analysis of 25 years, the ANN results show that the best hybrid solution for this case study is a grid + hydro with an NPV of €18,966, and the CES results point out for the same solution a NPV of €18,950, with a relative error of 0.08% and a correlation coefficient of 0.999996. Figure 16 presents the results for all configurations calculated by ANN and CES models showing clearly the best solution.

The negative value of NPV in Grid+Wind and Grid+Hydro+Wind is derived from initial installation costs of the wind turbine and its small energy production. For the case study a bigger wind turbine with a higher installed power capacity wasn't chosen because the wind speed in the case study area is very low and wind turbines that have a satisfactory energy production for these wind speeds are extremely expensive, being inadequate to the case study that is a small system and without many resources to be invested.

5. Conclusions

5.1. Optimization of the pumps' schedule in the Fátima system

The feasibility of the developed HGA model in the search of the best operational strategy for a lowest energy cost in the real Fátima system was analysed. Two algorithms were developed and linked to the GA. The first one, a repair algorithm from an analysis done in the unfeasible solutions generated by the GA, alters the decision variables in the attempt of making these solutions feasible. After finishing the generations of the GA, the second algorithm acts in these solutions, making a local search in the attempt of finding optimal locals.

The efficiency of the algorithm developed HGA in the search of the solution with lowest operational cost is confirmed, whereas the convergence occurred six times faster. One of the biggest limitations of the GA is the treatment of problems with high quantity of constraints. The application of penalties only allows the identification of unfeasible solutions. In problems of this kind it is probable that along the candidates' generation, the quantity of unfeasible candidates does not decrease, making the search of good solutions very difficult. With the implementation of repair algorithms, the appearance of super-candidates occurs in less time, since the alterations in the individuals are done directly in its problematic genes.

An evaluation analysis about the necessity of use genetic operators, when these algorithms are applied directly in a large set of solutions generated randomly, also shows final good results. To determine the best strategy among thousands of possible solutions it must also be taken into consideration the hydraulic reliability of the system.

The HGA model presented can be implemented in any network. Furthermore, its application is practical and useful, being able to be used by water supply companies, making easier the best decision aiming at the energy efficiency in pumping systems.

5.2. Prediction of hybrid energy solutions in Espite system

The current research work aims at the prediction analysis about the best energy system configuration, depending on the renewable available sources of the region, and the optimization of operating strategies for the water distribution systems (WDS), which have about 80% of their costs associated to the energy consumption. Hence an integrated methodology based on economical, technical and hydraulic performances has been developed using the following steps: (i) Artificial Neural Network (ANN) to determine the best hybrid energy system configuration; (ii) for the ANN training process, a configuration and economical base simulator model (CES) is used; (iii) as well a hydraulic and power simulator model (HPS) to describe the hydraulic behaviour; (iv) an optimization based-model to minimize pumping costs and maximize hydraulic reliability and energy efficiency is then applied.

The objective is to capture the knowledge domain in much more efficient way than a CES, ensuring a good reliability and best economical hybrid energy solution in the improvement of energy efficiency and sustainability of WDS. In this case study the installation of a micro hydro using water level controls and pump operation optimization shows the improvement of the energy efficiency in 63.35%. In this methodology to determine the best hybrid energy solution, the ANN has demonstrated significant reduction in time modelling, with a good a correlation and mean relative error.

Author details

H. M. Ramos[1*], L. H. M. Costa[2] and F. V. Gonçalves[3]

*Address all correspondence to: hr@civil.ist.utl.pt

1 Instituto Superior Técnico, Technical University of Lisbon, Portugal

2 Federal University of Ceará, Brazil

3 Federal University of Mato Grosso do Sul, Brazil

References

[1] Al-Alawi, A., M Al-Alawi, S., & M Islam, S.. (2007). Predictive control of an integrated PV-diesel water and power supply system using an artificial neural network. *Renewable Energy,,* 32(8), 1426-1439.

[2] Alanne, K., & Saari, A. (2006). Distributed energy generation and sustainable development. *Renewable and Sustainable Energy Reviews,,* 10(6), 539-558.

[3] Alvisi, S., Franchini, M., & Marinelli, A. (2007). A short-term, pattern-based water demand-forecasting model. *Journal of Hydroinformatics,,* 9(1), 39-50.

[4] Araujo, L. d. S. (2005). *Controlo de perdas na gestão sustentável dos sistemas de abastecimento de água.,* Lisbon: Instituto Superior Técnico, (Tese de Doutoramento), 338.

[5] Barsoum, N. N., & Vacent, P. (2007). Paper presented at First Asia International Conference on Modelling & Simulation- AMS'07. *Balancing Cost, Operation and Performance in Integrated Hydrogen Hybrid Energy System.,* 14-18.

[6]   Brion, L. M., & Mays, L. W. (1991). Methodology for Optimal Operation of Pumping Stations in Water Distribution Systems. *Journal of Hydraulic Engineering, ASCE.,* 117(11), 1551-1569.

[7]   Castronuovo, E. D., & Lopes, J. A. P. (2004). On the optimization of the daily operation of a wind-hydro power plant. *IEEE Transactions on Power Systems.,,* 19(3), 1599-1606.

[8]   Chaves, P., Tsukatani, T., & Kojiri, T. (2004). *Operation of storage reservoir for water quality by using optimization and artificial intelligence techniques Mathematics and Computers in Simulation,* 67(4-5), 419-432.

[9]   Christodoulou, S., & Deligianni, A. (2010). A Neurofuzzy Decision Framework for the Management of Water Distribution Networks. *Water Resources Management,,* 24(1), 139-156.

[10] Dihrab, S. S., & Sopian, K. (2010). Electricity generation of hybrid PV/wind systems in Iraq. *Renewable Energy,,* 35(6), 1303-1307.

[11] European Commission (2001). Directive for the promotion of electricity from renewable energy sources in the internal electricity market. In: Department of Energy of the European Commission, ed. 2001/77/EC Official Journal of the European Communities, , 33-40.

[12] Fabrizio, E., Corrado, V., & Filippi, M. (2010). A model to design and optimize multi-energy systems in buildings at the design concept stage. *Renewable Energy,* 35(3), 644-655.

[13] Fattahi, P., & Fayyaz, S. (2010). A Compromise Programming Model to Integrated Urban Water Management. *Water Resources Management,,* 24(6), 1211-1227.

[14] Francato, A. L., & Barbosa, P. S. F. (1999). Paper presented at XV Congresso Brasileiro de Engenharia Mecânica., Águas de Lindóia, SP. *Operação Ótima de Sistemas Urbanos de Abastecimento de Água.*

[15] Gabrovska, K., Wagner, A., & Mihailov, N. (2004). Software system for simulation of electric power processes in photovoltaic-hybrid system. *Proceedings of the 5th international conference on Computer systems and technologies: ACM,,* 1-7.

[16] Goldberg, D. E. (1989). Genetic Algorithm is Search Optimization and Machine Learning. Addision-Wesley-Longman, Reading.MA.

[17] Gupta, A., Saini, R. P., & Sharma, M. P. (2006). Modelling of Hybrid Energy System for Off Grid Electrification of Clusters of Villages. *Power Electronics, Drives and Energy Systems, 2006- PEDES'06.,,* 5.

[18] Hilton, A. B. C., & Culver, T. B. (2000). Constraint Handling for genetic algorithms in optimal remediation design. *J.Wat. Res. Plann. Mngmnt. ASCE,* 126(3), 128-137.

[19] Jafar, R., & Shahrour, I. (2007). Modelling the structural degradation in water distribution systems using the artificial neural networks (ANN). *Water Asset Management International,,* 3(3), 14-18.

[20] Jamieson, D. G., Shamir, U., Martinez, F., & Franchini, M. (2007). Conceptual design of a generic, real-time, near-optimal control system for water-distribution networks. *Journal of Hydroinformatics,,* 9(1), 3-14.

[21] Jamieson, D. G. (2007). Conceptual design of a generic, real-time, near-optimal control system for water-distribution networks. *Journal of Hydroinformatics,,* 9(1), 3-14.

[22] Jowitt, P. W., & Germanopoulos, G. (1992). Optimal Pump Scheduling in Water Supply Networks. *Journal of Water Resources Planning and Management, ASCE,,* 118(4), 406-422.

[23] Kenfack, J., Neirac, F. P., Tatietse, T. T., Mayer, D., Fogue, M., & Lejeune, A. (2009). Microhydro-PV-hybrid system: Sizing a small hydro-PV-hybrid system for rural electrification in developing countries. *Renewable Energy,,* 34(10), 2259-2263.

[24] Koroneos, C., Spachos, T., & Moussiopoulos, N. (2003). Exergy analysis of renewable energy sources. *Renewable Energy,,* 28(2), 295-310.

[25] Lansey, K. E., & Awumah, K. (1994). Optimal Pump Operations Considering Pump Switches. Journal of Water Resources Planning and Management, ASCE, ., 120(1), 17-35.

[26] Little, K. W., & Mc Crodden, B. J. (1989). Minimization of Raw Water Pumping Cost Using MILP. *Journal of Water Resources Planning and Management, ASCE,,* 115(4), 511-522.

[27] Martinez, F., Hernandez, V., Alonso, J. M., Rao, Z. ., & Alvisi, S. (2007). Optimizing of the operation of the Valencia water-distribution network. *J. Hydroinformatics,* 9(1), 65-78.2.

[28] Menegaki, A. (2008). Valuation for renewable energy: A comparative review. *Renewable and Sustainable Energy Reviews,,* 12(9), 2422-2437.

[29] Moura, P., & Almeida, A. d. (2009). Paper presented at Renewable Energy World Europe 2009., Cologne, Germany,. *Methodologies and Technologies for the Integrations of Renewable Resources in Portugal.,* 20.

[30] Ormsbee, L. E., Waski, T. M., Chase, D. V., & Sharp, W. W. (1989). Methodology for Improving Pump Operation Efficiency. *Journal of Water Resources Planning and Management, ASCE,,* 115(2), 148-164.

[31] Ormsbee, L. E., & Reddy, S. L. (1995). Nonlinear Heuristic for Pump Operations. *Journal of Water Resources Planning and Management, ASCE.,* 302-309.

[32] Ramos, H. M., Mello, M., & De , P. K. (2010). Clean power in water supply systems as a sustainable solution: from planning to practical implementation. *Water Science & Technology: Water Supply-WSTWS,,* 10(1), 39-49.

[33] Ramos, J. S., & Ramos, H. M. (2009a). Solar powered pumps to supply water for rural or isolated zones: A case study. . Energy for Sustainable Development , 13(3), 151-158.

[34] Ramos, J. S., & Ramos, H. M. (2009b). Sustainable application of renewable sources in water pumping systems: optimised energy system configuration. *Energy Policy,,* 37(2), 633-643.

[35] Rao, Z., & Alvarruiz, F. (2007). Use of an artificial neural network to capture the domain knowledge of a conventional hydraulic simulation model. *Journal of Hydroinformatics,,* 9(1), 15-24.

[36] Rao, Z., & Alvarruiz, F. (2007). Use of an artificial neural network to capture the domain knowledge of a conventional hydraulic simulation model. *Journal of Hydroinformatics,,* 9(1), 15-24.

[37] Rao, Z., & Salomons, E. (2007). Development of a real-time, near-optimal control system for water-distribution networks. *Journal of Hydroinformatics,,* 9(1), 25-38.

[38] Rossman, L. A. (2000). *EPANET User's Manual. US EPA. Cincinnati,OH.*

[39] Sakarya, A. B. A., & Mays, L. W. (2000). Optimal Operation of Water Distribution Pumps Considering Water Quality. *Journal of Water Resources Planning and Management,,* 126(4), 210-220.

[40] Salomons, E., Goryashko, A., Shamir, U., Rao, Z., & Alvisi, S. (2007). Optimizing the operation of the Haifa-A water-distribution network,. *J. Hydroinformatics,* 9(1), 51-64.

[41] Setiawan, A. A., Zhao, Y., & Nayar, C. V. (2009). Design, economic analysis and environmental considerations of mini-grid hybrid power system with reverse osmosis desalination plant for remote areas. *Renewable Energy,,* 34(2), 374-383.

[42] Shamir, U., & Salomons, E. (2008). Optimal Real-Time Operation of Urban Water Distribution Systems Using Reduced Models. *ASCE,* 181-185.

[43] Turgeon, A. (2005). Daily Operation of Reservoir Subject to Yearly Probabilistic Constraints. *Journal of Water Resources Planning and Management,,* 131(5), 342-350.

[44] Vieira, F., & Ramos, H. M. (2008). Hybrid solution and pump-storage optimization in water supply system efficiency: A case study. *Energy Policy,,* 36(11), 4142-4148.

[45] Vieira, F., & Ramos, H. M. (2009). Optimization of operational planning for wind/ hydro hybrid water supply systems. *Renewable Energy,* [3], 34, 928-936.

[46] Watergy (2009). http://www.watergy.net/-overview/ why.php. Acessed in: 09/28/2009.

[47] Wood, D. J., & Reddy, L. S. (1994). Control de Bombas de Velocidad Variable y Modelos en Tiempo Real para Minimizar Fugas y Costes Energéticos,. In: Mejora del Rendimiento y de La Fiabilidad en Sistemas de Distribucion de Agua. València, Espanha: Editores E. Cabrera, U. D. Mecánica de Fluidos, Universidad Politécnica de Valencia, A. F. Vela e Universitat Jaume I de Castellón, , 173-207.

Chapter 5

Water Demand Uncertainty: The Scaling Laws Approach

Ina Vertommen, Roberto Magini,
Maria da Conceição Cunha and Roberto Guercio

Additional information is available at the end of the chapter

1. Introduction

Water is essential to all forms of life. The development of humanity is associated to the use of water, and nowadays, the constant availability and satisfaction of water demand is a fundamental requirement in modern societies. Although water seems to be abundant on our planet, fresh water is not an inexhaustible resource and has to be managed in a rational and sustainable way. The demand for water is dynamic and influenced by various factors, from geographic, climatic and socioeconomic conditions, to cultural habits. Even within the same neighbourhood the user-specific water demand is elastic to price, condition of the water distribution system (WDS), air temperature, precipitation, and housing composition (regarding only residential demand in this case). On top of all these factors, demand varies during the day and the week.

Traditionally, for WDS modelling purposes, water demand is considered as being deterministic. This simplification worked relatively well in the past, since the major part of the studies on water demands were conducted only with the objective of quantifying global demands, both on the present and on the long-term. With the development of optimal operating schedules of supply systems, hourly water demand forecasting started to become increasingly more important. Moreover, taking in consideration all the aforementioned factors that influence water use, it is clear that demand is not deterministic, but stochastic. Thus, more recently, in order to guarantee the requested water quantities with adequate pressure and quality, the studies began to focus on instantaneous demands and their stochastic structure.

1.1. Descriptive and Predictive Models for Water Demand

The first stochastic model for (indoor) residential water demands was proposed by Buchberger and Wu [1]. According to the authors, residential water demand can be characterized by three parameters: frequency, duration and intensity, which in turn can be described by a Poisson rectangular pulse process (PRP). The adopted conceptual approach is relatively similar to basic notions of queuing theory: a busy server draws water from the system at a random, but constant, intensity and, during a random period of time. Residential demands were subdivided into deterministic and stochastic servers. Deterministic servers, including washing machines and toilets, produce pulses which are always similar. Stochastic servers, like water taps, instead produce pulses with great variability, and their duration and intensity are independent. The PRP process found to best describe water demand is non-homogeneous, i.e., when the pulse frequency is not constant in time. Different authors used real demand data to assess the adequacy of the non-homogeneous PRP model, achieving good results [2]. Moreover, the PRP model was confirmed to allow the characterization of the spatial and temporal instantaneous variability of flows in a network, unlike the traditional models that use spatial and temporal averages and neglect the instantaneous variations of demand. One drawback to the rectangular pulse based models is the fact that the total intensity is not exactly equal to the sum of the individual intensities of overlapping pulses, due to increased head loss caused by the increased flow [3]. This problem can however be solved by introducing a correction factor. The daily variability of demand represents another drawback to the PRP model, since it can invalidate the hypothesis that pulses arrive following a time dependent Poisson process [2]. One possible solution to this question is to treat the time dependent non-homogeneous process as a piecewise homogeneous process, by dividing the day into homogeneous intervals [4]. Another solution consists in using an alternative demand model: the cluster Neyman-Scott rectangular pulse model (NSRP), proposed by Alvisi [5]. The model is similar to the PRP model, but the total demand and the frequency of pulses are obtained in different ways. In the PRP model the total water demand follows a Poisson process resulting from the sum of the single-user Poisson processes, with a single arrival rate. In the NSRP model, a random number of individual demands (or elementary demands) are aggregated in demand blocks. The origin of the demand blocks is given by a Poisson process, with a certain rate between the subsequent arrivals. The temporal distance between the origins of each of the elementary demands to the origin of the demand block, follows an exponential distribution with a different rate. The variation of these parameters during the day reflects the cyclic nature of demands. A good approximation of the statistical moments for different levels of spatial and temporal aggregation was achieved; however, the variance of demand becomes underestimated for higher levels of spatial aggregation.

The aforementioned models are mainly descriptive. More recently, Blokker and Vreeburg [6] developed a predictive end-use model, based on statistical information about users and end uses, which is able to forecast water demand patterns with small temporal and spatial scales. In this model, each end-use is simulated as a rectangular pulse with specific probability distribution functions for the intensity, duration and frequency, and a given probability of use over the day. End-uses are discriminated into different types (bath, bathroom tap, dish washer, kitchen tap, shower, outside tap, washing machine, WC). The statistical distribution for the frequency of each end-use was retrieved from survey information from the Netherlands. The duration and intensity were determined, partly from the survey and partly from technical information on water-using appliances. From the retrieved information, a diurnal pattern could be built for each user. Users represent a key point in the model and are divided into groups based on household size, age, gender and occupation. Simulation results were found to be in good agreement with measured demand data. The End-Use model has also been combined with a network solver, obtaining good results for the travel times, maximum flows, velocities and pressures [7].

The PRP and the End-Use model were compared against data from Milford, Ohio. The achieved results showed that both models compare well with the measurements. The End-Use model performs better when simulating the demand patterns of a single family residence, while the PRP models describes more accurately the demand pattern of several aggregated residences [8]. The main difference between the models is the number of parameters they use: the PRP model is a relatively simple model that has only a few parameters, while the End-Use model has a large number of parameters. However, the End-Use model is very flexible towards the input parameters, which also have a clearer physical meaning and hence more intuitive to calibrate. The PRP model describes the measured flows very well. From the analytic description provided by the PRP model, a lot of mathematical deductions can be made. Thus, one can classify the PRP model as a descriptive model with a lot of potential to provide insight into some basic elements of water use, such as peak demands [9]and cross-correlations [10]. The End-Use model is a Monte Carlo type simulation that can be used as a predictive model, since it produces very realistic demand patterns. The End-Use model can be applied in scenario studies to show the result of changes in water using appliances and human behaviour. Possible improvements to the model include the incorporation of leakage, the consideration of demands as a function of the network pressure and the application of the model outside the Netherlands [11]. Li [10]studied the spatial correlation of demand series that follow PRP processes. It was verified that while time averaged demands that follow a homogeneous PRP process are uncorrelated, demands that follow a non-homogenous PRP process are correlated, and that this correlation increases with spatial and temporal aggregation. A similar conclusion about the correlation was achieved by Moughton [12]from field measurements.

1.2. Uncertainty and reliability-based design of water distribution systems

The problem of WDS design consists in the definition of improvement decisions that can optimize the system given certain objectives. As aforementioned, in the earliest works regarding the optimal design of water distribution systems (WDS), input parameters, like water demand, were considered as being deterministic, often leading to under-designed networks. A robust design, allowing a system to remain feasible under a variety of values that the uncertain input parameters can assume, can only be achieved through a probabilistic approach. In a probabilistic analysis the input parameters are considered to be random variables, i.e., the single values of the parameters are replaced with statistical information that illustrates the degree of uncertainty about the true value of the parameter. The outcomes, like nodal heads, are consequently also random variables, allowing the expression of the networks' reliability.

Uncertainty in demand and pressure heads was first explicitly considered by Lansey [13]. The authors developed a single-objective chance constrained minimization problem, which was solved using the generalized reduced gradient method GRG2. The obtained results showed that higher reliability requirements were associated to higher design costs when one of the variables of the problem was uncertain.

Xu and Goulter [14] proposed an alternative method for assessing reliability in WDS. The mean values of pressure heads were obtained from the deterministic solution of the network model. The variance values were obtained using the first-order second moment method (FOSM). The probability density function (PDF) of nodal heads defined by these mean and variance values was used to estimate the reliability at each node. The approach proved to be suitable for demands with small variability. Kapelan [15] developed two new methods for the robust design of WDS: the integration method and the sampling method. The integration method consists in replacing the stochastic target robustness constraint (minimum pressure head) with a set of deterministic constraints. For that matter it is necessary to know the mean and standard deviation of the pressure heads. However, since pressure heads are dependent of the demands, it is not possible to obtain analytically the values for the standard deviations. Approximations of the values of the standard deviations are obtained by assuming the superposition principle, which makes it possible to estimate the contribution of the uncertainty in demand on the uncertainty of pressure heads. The sampling method is based on a general stochastic optimization framework, this is, a double looped process consisting on a sampling loop within an optimization loop. The optimization loop finds the optimal solution, and the sampling loop propagates the uncertainty in the input variables to the output variables, thus evaluating the potential solutions.

The aforementioned optimization problems are formulated as constrained single-objective problems, resulting in only one optimal solution (minimum cost), that provides a certain level of reliability. More recently, these optimization problems have been replaced with multi-objective problems. Babayan [16] formulated a multi-objective optimization problem considering two objectives at the same time: the minimization of the design cost and the maximization of the systems' robustness. Nodal demands and pipe roughness coefficients were assumed to be independent random variables following some PDF.

At this point, all the aforementioned models assume nodal demands as independent random variables. However, in real-life demands are most likely correlated: demands may rise and fall due to the same causes. Kapelan [17] introduced nodal demands as correlated random variables into a multi-objective optimization problem. The authors verified that the optimal design solution is more expensive when demands are correlated than the equivalent solution when demands are uncorrelated. A similar conclusion was achieved by Filion [18]. These results sustain that assuming uncorrelated demands can lead to less reliable network designs. Thus, even if increasing the complexity of optimization problems, demand correlation should always be taken into account in the design of WDS.

The robust design of WDS has gained popularity over the last years. Researchers have been focusing on methods and algorithms to solve the stochastic optimization problems, and great improvements have been made in this aspect. However, the quantification of the uncertainty itself has not been addressed. Values for the variance and correlation of nodal demands are always assumed and no attention is being paid in properly quantifying these parameters. The optimization problems could be significantly improved if more realistic values for the uncertainty would be taken into account.

This work addresses the need to understand in which measure the statistical parameters depend on the number of aggregated users and on the temporal resolution in which they are estimated. It intends to describe these dependencies through scaling laws, in order to derive the statistical properties of the total demand of a group of users from the features (mean, variance and correlation) of the demand process of a single-user. Being part of the first author's PhD research, which aims the development of descriptive and predictive models for water demand that provide insight into peak demands, extreme events and correlations at different spatial and temporal scales, these models will, in future stages, be incorporated in decision models for design purpose or scenario evaluation. Through this approach, we hope to develop more realistic and reliable WDS design and management solutions.

2. Statistical characterization of water demand

Recent studies on uncertainty in water distribution systems (WDS) refer that nodal demands are the most significant inputs in hydraulic and water quality models [19]. The variability of water demand affects the overall reliability of the model, the assessment of the spatial and temporal distributions of the pressure heads, and the evaluation of water quality along the different pipes. These uncertainties assume a different importance depending on the spatial and temporal scales that are considered when describing the network. The degree of uncertainty becomes more relevant when finer scales are reached, i.e., when small groups of users and instantaneous demands are considered. Thus, for a correct and realistic design and management, as well as simulation and performance assessment of WDS it is essential to have accurate values of water demand that take into account the variability of consumption at different scales. For that matter, the thorough description of the statistical properties of demand of the different groups of customers in the network, at specific temporal resolutions, is essential.

For a better understanding of this aspect, let us consider the distribution of the customers in a network. Figure 1 shows the network of a small town where the customers can be classified mainly as residential.

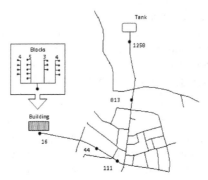

Figure 1. Spatial variability of customers in a real distribution network, from Magini *et al*. [25]. The number of customers is outlined at various locations.

The most peripheral pipe serves the inhabitants of one single building. When moving upwards in the network, the number of customers increases reaching a maximum of 1258 customers near the tank. Obviously, as a consequence, the mean flow increases from the peripheral building to the tank. The increase of the variance of the flow is, however, less obvious. For larger networks and more densely populated towns, the difference between the number of customers that are close and far from the tank, and consequently, the variations of the mean and variance of the flow is even more pronounced.

Another important aspect when modelling a network is the choice of the adequate temporal resolution. This choice depends on the characteristics of the available measurement instruments and on the type of analysis to perform. When modelling the peripheral part of a network, characterized by a significant temporal variation of demand, it is important to adopt fine temporal resolutions, i.e., in the order of seconds. For the estimation of peak flows in design problems Tessendorff [20] suggests the use of different temporal resolutions on different sections of the network: the author suggests the use of a 15 second time interval for customer installation lines, two minutes for service lines, 15 minutes for distribution lines, and 30 minutes for mains and secondary feeders. The statistical properties of water demand are affected by the considered temporal resolution. The use of longer sampling intervals causes an inevitable loss of information about the signals, resulting in lower estimates for the variance [21, 22]. This aspect is particularly relevant at the peripheral pipes of the network that, as aforementioned, are characterized by large demand fluctuations. Therefore, understanding the spatial and temporal scaling properties of water demand is essential to build a stochastic model for water consumption.

Water demand can be described by a stochastic process in which $q(i, t)$ represents the demand of water of the single-user i at time instant t. In order to estimate the statistical properties of water demand, it is necessary to have a historical series of observations, extended to sufficiently wide number of users of each type. From this data it is then possible to estimate the mean and variance of the process.

If the consumers are assumed to be of the same type, the properties of demand can be considered to be homogeneous in space, this is, they are independent of the particular consumer that is taken into consideration. Regarding the temporal variability, the stochastic process can only be assumed to be stationary in time intervals during which the mean stays constant. Once the length of this time interval, T, is established, it is possible to determine the temporal mean, μ_1, and variance, σ_1^2, of the demand signal of the single-user i, as followed:

$$\mu_1 = \frac{1}{T}\int_0^T q(i, t)dt \tag{1}$$

$$\sigma_1^2 = \frac{1}{T}\int_0^T [q(i, t) - \mu_1]^2 dt \tag{2}$$

For homogeneous and stationary demands, the expected values for the mean and variance, $E[\mu_1]$ and $E[\sigma_1^2]$, obtained from N observations, provide the mean and variance of the process.

2.1. Correlation between consumers

The definition of the mean and variance for each type of consumer is not enough for a complete statistical characterization of demand. In order to obtain a realistic representation of the demand loads at the different nodes in a network; essential for the assessment of the network performance under conditions as close as possible to the actual working conditions, the correlation between nodal demands cannot be ignored. This correlation can be expressed through the cross-covariance and cross-correlation coefficient functions.

The cross-covariance, cov_{AB}, and cross-correlation coefficient, ρ_{AB}, between user i of group A and user j of group B, during the observation period T, are expressed, respectively, as followed:

$$cov_{AB} = \frac{1}{T}\int_0^T [q_A(i_A, t) - \mu_A][q_B(j_B, t) - \mu_B]dt \tag{3}$$

$$\rho_{AB} = \frac{cov_{AB}}{\sigma_A \cdot \sigma_B} \tag{4}$$

As known, the WDS need to guarantee minimum working conditions, this is, the minimum pressure requirements have to be satisfied at each node even under maximum demand loading conditions. If all the consumers in the network are of the same type, it seems reasonable to assume a perfect correlation between demands, and to simplify the analysis of the network by assigning the same demand pattern to all the consumers. The synchronism of demands is the worst scenario that can occur on a network, causing the widest pressure fluctuations at the nodes. The assumption of a perfect correlation for design purposes results in reliable networks, but it also requires the increase of the pipe diameters, which consequently increases the networks cost. In fact, as mentioned earlier, each consumer has his own demand pattern based on specific needs and habits, without knowing what other consumers are doing at the same time. This means that demand signals in real networks are correlated, but are not synchronous. Thus, in order to obtain the optimum design of a network, it is essential to estimate the accurate level of correlation between the consumers. On the other hand, to estimate accurately the spatial correlation between demands, it is necessary to collect and analyse historical series, resulting in additional costs in the design phase. However, these additional costs will most certainly be compensated by the achieved reduction of the following construction costs.

2.2. The scaling laws approach in modelling water demand uncertainty

Water demand uncertainty is made of both aleatory or inherent uncertainty, due to the natural and unpredictable variability of demand in space and time, and epistemic or internal uncertainty, due to a lack of knowledge about it. Hutton [23] distinguishes epistemic uncertainty in two types. The first type concerns the nature of the demand patterns, and the lack of knowledge about this variability when modelling WDS both in time and space. This uncertainty is defined as 'two-dimensional' uncertainty since it is composed by both aleatory and epistemic uncertainty. It can be reduced with extended and expensive spatial and temporal data collection or through the employment of descriptive and predictive water demand models. The second type of epistemic uncertainty takes the spatial allocation of water demand into account when modelling WDS [24].

Dealing with the 'two-dimensional' uncertainty when modelling WDS, requires not only a complete statistical characterization of demand variability, but also the determination of the correlation among the different users and groups of users. The natural variability of demand can be expressed using probability density functions (PDF). A PDF is characterized by its shape (e.g. normal, exponential, gamma, among others) and by specific parameters like the population mean and variance. Thus, in order to represent uncertain water demand using a PDF, it is necessary to identify and estimate the values of these parameters. The consideration of different spatial and temporal aggregation levels induces changes in the PDF parameters, often leading to a reduction of the uncertainty. The auto-correlation and cross-correlation that characterize the water demand signals affect the extent to which the PDF parameters vary, and can introduce an additional sensitivity to the specific period of observation in question.

In order to understand the effects of spatial aggregation and sampling intervals on the statistical properties of demand, it is possible to develop analytical expressions for the moments (mean, variance, cross-covariance and cross-correlation coefficient) of demand time series, at a fixed time sampling frequency Δt, of n aggregated users as a function of the moments of the single-user series sampled in the observation period T. These expressions are referred to as "*Scaling Laws*", and can be expressed as:

$$E[m_{\Delta t,T}(n)] = E[m_{\Delta t,T}] \cdot n^{\alpha} \cdot f(\Delta t, T) \tag{5}$$

Where $E[m_{\Delta t,T}(n)]$ is the expected value of the moment m for n users for the time interval T; $E[m_{\Delta t,T}]$ is the expected value of the moment m for the single-user for the same time interval; a is the exponent of the scaling law; and $f(\Delta t, T)$ is a function that expresses the influence of both sampling rate and observation period.

The development of the scaling laws is based on the assumption that the demand can be described by a homogeneous and stationary process, which implies that the n aggregated users are of the same type (residential, commercial, industrial, etc.), and that the statistical properties of demand, mean and variance, can be assumed constant in time. The scaling laws for the mean, variance, and lag1 covariance were derived by Magini [25]. The expected value of the total mean demand $q(n, t)$ can be expressed as followed:

$$E[\mu_{\Delta t,T}(n)] = E\left[\frac{1}{T}\int_0^T \sum_{j=1}^n q(j, \tau)d\tau\right] = \frac{1}{T}\int_0^T \sum_{j=1}^n E[q(j, \tau)]d\tau = n \cdot E[\mu_1] \tag{6}$$

Where $E[\mu_1]$ is the expected demand value for the single user or 'unit mean'. This expression shows that the mean demand increases linearly with the number of users according to a factor of proportionality equal to the expected value of the single user and is independent of the sampling rate and observation period.

In order to estimate the expected value of the demand variance it is necessary to consider the covariance function $cov(s, \tau)$ of the single-user demand at the spatial and temporal lags, $s = j_1 - j_2$ and $\tau = \tau_1 - \tau_2$, respectively. The following expression is obtained (see [26] for the mathematical passages):

$$E[\sigma^2_{\Delta t,T}(n)] = \frac{1}{T^2}\int_0^T\int_0^T \sum_{i_1=1}^n \sum_{i_2=1}^n [cov_{\Delta t}(s,0) - cov_{\Delta t}(s, \tau)]d\tau_1 d\tau_2$$
$$= \sum_{i_1=1}^n \sum_{i_2=1}^n \left[cov_{\Delta t}(s,0) - \frac{1}{T^2}\int_0^T\int_0^T cov_{\Delta t}(s, \tau)d\tau_1 d\tau_2\right] \tag{7}$$

Where $cov_{\Delta t}(s, 0)$ is the covariance function at lag $s = 0$, and $cov_{\Delta t}(s, \tau)$ is the space-time covariance function. This expression shows that the expected value for the sample variance of the n-users process depends on the correlation structure of the single-user demands. The term $\frac{1}{T^2}\int_0^T\int_0^T cov_{\Delta t}(s, \tau)d\tau_1 d\tau_2$ decreases as the period of observation T increases, becoming negligible when $T >> \theta$, being θ a parameter, connected to the cross-correlation of the demands and similar to the scale of fluctuation for the auto-correlation of a single signal.

The term $cov_{\Delta t}(s, 0)$ is independent from τ_1 and τ_2, and assumes the following values:

$$cov_{\Delta t}(s,0) = \begin{cases} cov_{\Delta t}(s) & j_1 \neq j_2 \\ \sigma^2_{1,\Delta t} & j_1 = j_2 \end{cases} \tag{8}$$

Where $cov_{\Delta t}(s)$ is the spatial cross-covariance between different single-user demands, and $\sigma^2_{1,\Delta t}$ the variance of the single user. For large values of T, equation (7) can be simplified into:

$$E\left[\sigma^2_{\Delta t,T}(n)\right]=\sum_{j_1=1}^{n}\sum_{j_2=1}^{n}cov_{\Delta t}(s,0)=\sum_{j_1=1}^{n}\sum_{j_2=j_1}^{n}cov_{\Delta t}(s,0)+\sum_{j_1=1}^{n}\sum_{j_2\neq j_1}^{n}cov_{\Delta t}(s,0)$$

$$=n\cdot\sigma^2_{1,\Delta t}+n\cdot(n-1)\cdot cov_{\Delta t}(s) \tag{9}$$

This equation represents the scaling law for the variance, neglecting the bias that can be caused when using small the demand series (short observation periods).

Introducing the Pearson cross-correlation coefficient given by $\rho=\frac{cov_{xy}}{\sigma_x\sigma_y}$, and considering $\rho_{\Delta t}$ as the cross-correlation coefficient between each couple of single-user demands, the spatial covariance can be expressed as $cov_{\Delta t}=\rho_{\Delta T}\sigma^2_{1,\Delta t}$, and Equation (9) becomes:

$$E\left[\sigma^2_{\Delta t}(n)\right]=n^2\cdot\rho_{\Delta t}\cdot\sigma^2_{1,\Delta t}+n\cdot\left[1-\rho_{\Delta t}\right]\cdot\sigma^2_{1,\Delta t} \tag{10}$$

If demands are perfectly correlated in space then $\rho_{\Delta t}$ is equal to one, and equation (10) is simplified into:

$$E\left[\sigma^2_{\Delta t}(n)\right]=n^2\cdot\sigma^2_{1,\Delta t} \tag{11}$$

If demands are uncorrelated in space then $\rho_{\Delta t}$ is equal to zero, and equation (9) is simplified into:

$$E\left[\sigma^2_{\Delta t}(n)\right]=n\cdot\sigma^2_{1,\Delta t} \tag{12}$$

Since the cross-correlation coefficient can assume values between 0 and 1, equations (10) and (11) represent the maximum and minimum expected values for the variance. Equation (9) can be simplified into a more generic form given by:

$$E\left[\sigma^2_{\Delta t}(n)\right]=n^\alpha\cdot\sigma^2_{1,\Delta t} \tag{13}$$

Where $1\leq\alpha\leq2$.

In conclusion, it can be stated that the variance in the consumption signal of a group of users n, homogeneous in type, is proportional to the mean variance of the single-user according to an exponent, which varies between 1 and 2. The value of the scaling exponent depends on the structure of the spatial correlation, i.e., the correlation that exists between the different consumptions during the observation period: if demands are uncorrelated in space, the scaling law is linear, if demands are perfectly correlated in space, the scaling law is quadratic.

The variance function, $\gamma(\Delta t)$, measures the reduction of the variance of the instantaneous signal when the sampling interval Δt increases [27], as followed:

$$\sigma_{1,\Delta t}^2 = \sigma_1^2 \cdot \gamma(\Delta t) \qquad (14)$$

Where σ_1^2 is the variance of the instantaneous signal for the single user. Introducing the variance function in equation (13), the following is obtained:

$$E\left[\sigma_{\Delta t}^2(n)\right] = n^\alpha \cdot \sigma_1^2 \cdot \varphi(\Delta t) \qquad (15)$$

Similarly, the expected value of the cross-covariance is given by:

$$
\begin{aligned}
E\left[cov_{AB,\Delta t}(n_a, n_b)\right] &= \frac{1}{T^2}\int_0^T\int_0^T\sum_{i=1}^{n_a}\sum_{j=1}^{n_b}\left[cov_{AB,\Delta t}(s,0) - cov_{AB,\Delta t}(s,\tau)\right]d\tau_1 d\tau_2 \\
&= \sum_{i=1}^{n}\sum_{j=1}^{n}\left[cov_{AB,\Delta t}(s,0) - \frac{1}{T^2}\int_0^T\int_0^T cov_{AB,\Delta t}(s,\tau)d\tau_1 d\tau_2\right]
\end{aligned} \qquad (16)
$$

Neglecting the term $\frac{1}{T^2}\int_0^T\int_0^T cov_{AB,\Delta t}(s,\tau)d\tau_1 d\tau_2$, the expected value of the cross-covariance between the demands of n_a aggregated users of group A and n_b aggregated users of group B is given by:

$$E\left[cov_{AB,\Delta t}(n_a, n_b)\right] = n_a \cdot n_b \cdot E\left[\rho_{ab,\Delta t}\right] \cdot E\left[\sigma_{a,\Delta t}\right] \cdot E\left[\sigma_{b,\Delta t}\right] \qquad (17)$$

Where, $E\left[\rho_{ab,\Delta t}\right]$ is the expected Pearson cross-correlation coefficient between the single-user demands of the two groups; and $\sigma_{a,\Delta t}$ and $\sigma_{b,\Delta t}$ are the standard deviations of the single-user demands of groups A and B, respectively, at the sampling rate Δt. The expected value of the cross-covariance increases according to the product between the number of users of each group. In the particular case in which both groups have the same statistical properties, i.e., they belong to the same process, and assuming that $n_a = n_b$, the scaling law of the cross-covariance becomes quadratic.

As a consequence, the expected value of the Pearson cross-correlation coefficient between the demands of n_a aggregated users of group A and n_b aggregated users of group B, is given by:

$$E\left[\rho_{AB,\Delta t}(n_a, n_b)\right] = \frac{E\left[Cov_{AB,\Delta t}(n_a, n_b)\right]}{E\left[\sigma_{A,\Delta t}(n_a)\right] \cdot E\left[\sigma_{B,\Delta t}(n_b)\right]} = \frac{n_a \cdot n_b \cdot E\left[\rho_{ab,\Delta t}\right]}{\sqrt{n_a\left(1 + E\left[\rho_{a,\Delta t}\right] \cdot \left[n_a - 1\right]\right)} \cdot \sqrt{n_b\left(1 + E\left[\rho_{b,\Delta t}\right] \cdot \left[n_b - 1\right]\right)}} \qquad (18)$$

This equation shows that this coefficient depends separately on the spatial aggregation levels of each group, n_a and n_b, and not only on their product as happens for the cross-covariance. If $n_a = n_b = n$ equation [18] becomes:

$$E\left[\rho_{AB,\Delta t}(n)\right] = \frac{n \cdot E\left[\rho_{ab,\Delta t}\right]}{\sqrt{\left[1 + (n-1) \cdot E\left[\rho_{a,\Delta t}\right]\right]\left[1 + (n-1) \cdot E\left[\rho_{b,\Delta t}\right]\right]}} \tag{19}$$

From equation (19) it is possible to observe that the expected value $E\left[\rho_{AB,\Delta t}(n_a, n_b)\right]$ increases with the number of users, n_a and n_b, reaching the following limit value:

$$E\left[\rho_{AB,\Delta t}(n_a, n_b)\right] = \lim_{\substack{n_a \to \infty \\ n_b \to \infty}} E\left[\rho_{AB,\Delta t}(n_a, n_b)\right] = \frac{E\left[\rho_{ab,\Delta t}\right]}{\sqrt{\left(E\left[\rho_{a,\Delta t}\right] \cdot E\left[\rho_{b,\Delta t}\right]\right)}} \tag{20}$$

Since by definition $E\left[\rho_{ab,\Delta t}\right] \le 1$, the maximum value that the expected value of the cross-correlation coefficient between the single-user demands of group A and B can assume is:

$$E\left[\rho_{AB,\Delta t}\right]_{max} = \sqrt{\left(E\left[\rho_{a,\Delta t}\right] \cdot E\left[\rho_{b,\Delta t}\right]\right)} \tag{21}$$

From equation (21) it is also possible to observe that the Pearson cross-correlation coefficient between the n_a aggregated users of group A and the n_b aggregated users of group B depends on both the cross-correlations inside each group and the cross-correlation between the groups. Therefore, it seems interesting to investigate the way in which these two aspects, one at a time, affect the expected value of the cross-correlation when the number of aggregated users increases and for a fixed sampling rate Δt. In order to do so let us first consider a fixed value of ρ_{ab} and varying values of ρ_a and ρ_b. Figure 2 shows the graphical results for $\rho_{ab} = 0.1$ and different pairs of ρ_a and ρ_b.

As expected, all the curves have a common starting point, since ρ_{ab} is fixed. According to equation (19) a gradual flattening of the curves and a reduction of the shape ratio $\rho_{AB,lim} / \rho_{ab}$ can be noticed when the product $\rho_a \cdot \rho_b$ increases. Let us now consider a different case in which ρ_a and ρ_b are fixed and ρ_{ab} varies. The results are shown graphically in figure 3. The curves have now different starting points and equal shape ratios $\rho_{AB,lim} / \rho_{ab}$. Increasing ρ_{ab} produces only an upward shift of the curves, extending their transient.

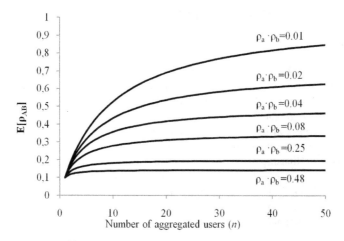

Figure 2. Scaling laws of $E[\rho_{AB}(n)]$, for different values of $\rho_a \cdot \rho_b$.

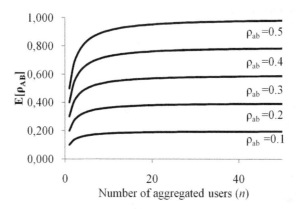

Figure 3. Scaling laws of $E[\rho_{AB}(n)]$, for different values of ρ_{ab}.

In the particular case in which both groups of users have the same statistical properties, i.e., they belong to the same process, and assuming $n_a = n_b = n$, the scaling law for the cross-correlation coefficient, considering no differences in the sampling time intervals, is:

$$E\left[\rho_{AB,\Delta t}(n)\right] = \frac{n \cdot E\left[\rho_{\Delta t}\right]}{1 + (n-1) \cdot E\left[\rho_{\Delta t}\right]} \tag{22}$$

From equation (22) it is clear that the cross-correlation coefficientincreases with the number of aggregated users, tending to one. This limit value is reached as sooner as the cross-correlation coefficient, $E\left[\rho_{\Delta t}\right]$, between the single-user demands is higher.

3. Validation of the Analytical expressions

3.1. Synthetically generated signals: scaling laws for the mean and the variance

In order to confirm the analytical development reported in the previous paragraph, the scaling laws were derived for groups of synthetically and simultaneously generated consumption signals. At this aim the Multivariate Streamflow model [28], with a normal probability distribution, was used. Each group was assumed to contain 300 consumption signals with 3600 realizations each, distinguished by different values of the cross-correlation coefficient between them. The correctness of the procedure used to generate each demand series was tested by checking that the mean, the variance and the cross-correlation coefficient of the generated signals were equal to the input parameters of the model. Only little differences were observed (Table 1), which are explained due to the fact that the generated demand series are realizations of a stochastic process and, consequently, their moments necessarily differ from the theoretical ones.

Once the single consumption signals of each group were generated, they were aggregated randomly selecting one at a time, until a maximum of 100 aggregated consumption signals was reached. The first and second order moments, mean and variance, were calculated for each aggregation level. In order to obtain a result as general as possible, the same procedure has been repeated 50 times, aggregating each time different users [25]. The obtained results are summarized in Table 1 and 2, with reference to equation 5.

Cross-correlation	$E[m_T]$		a	
coefficient	Theoretical	Experimental	Theoretical	Experimental
0	0.70	0.7003	1.00	0.9996
0.001	0.70	0.7017	1.00	0.9993
0.010	0.70	0.6971	1.00	1.0001
0.025	0.70	0.7020	1.00	0.9989
0.050	0.70	0.7096	1.00	1.0004
0.10	0.70	0.7063	1.00	1.0009
0.20	0.70	0.7086	1.00	0.9994
0.30	0.70	0.7032	1.00	1.0008
0.40	0.70	0.6923	1.00	1.0003
0.50	0.70	0.6942	1.00	0.9985
0.60	0.70	0.6857	1.00	1.0002
0.70	0.70	0.6874	1.00	1.0011
0.80	0.70	0.6852	1.00	0.9998
0.90	0.70	0.6789	1.00	1.0009
0.99	0.70	0.7050	1.00	0.9997

Table 1. Theoretical and experimental values of the scaling law for the first order moment for different values of the cross-correlation.

Cross-correlation coefficient	$E[m_T]$	a
0	3.9808	1.0004
0.001	3.5205	1.0541
0.010	1.9864	1.3079
0.025	1.4498	1.4984
0.050	1.2702	1.6403
0.10	1.2940	1.7686
0.20	1.5621	1.8675
0.30	1.8879	1.9139
0.40	2.1713	1.9379
0.50	2.4485	1.9570
0.60	2.7685	1.9692
0.70	3.0803	1.9804
0.80	3.4051	1.9888
0.90	3.6567	1.9945
0.99	3.9960	1.9985

Table 2. Experimental values of the scaling law for the second order moment for different values of the cross-correlation.

Results confirm the linear scaling for the first order moment and show that the variance increases with the spatial aggregation level according to an exponent that varies between 1 and 2. In theory, for spatially uncorrelated demands the scaling laws is linear and for perfectly correlated demands the scaling law is quadratic.

3.2. Synthetically generated signals: scaling laws for the cross-covariance

In this case pairs of aggregated consumption series, A and B, were obtained by randomly selecting among pairs of the previously generated groups of signals. Different values of the product $n_a \cdot n_b$, where n_a is the number of signals in group A and n_b the same number in group B, were considered, up to the maximum value $n_a \cdot n_b = 500$. Each aggregation process was characterized by the cross-correlation value between the single signals in the same group and the cross-correlation value between the single signals of the two native groups. The cross-covariance was computed for the different aggregation levels and the scaling law were derived for each process. The results are summarized in Table 3 with reference to equation 17, considering $Coeff = E[\rho_{ab,\Delta T}] \cdot E[\sigma_{a,\Delta T}] \cdot E[\sigma_{b,\Delta T}]$ and α as the exponent of the product $n_a \cdot n_b$.

Cross-correlation coefficient			Coeff		α	
ρ_a	ρ_b	ρ_{ab}	theoretical	experimental	theoretical	experimental
0.10	0.10	0.10	0.3754	0.3747	1.00	0.9998
0.20	0.10	0.10	0.3880	0.3907	1.00	0.9982
0.40	0.10	0.10	0.3676	0.3624	1.00	1.0020
0.40	0.20	0.10	0.3463	0.3440	1.00	1.0003
0.80	0.60	0.10	0.3939	0.3908	1.00	1.0009
0.50	0.50	0.10	0.3545	0.3541	1.00	0.9993
0.50	0.50	0.20	0.7449	0.7401	1.00	1.0007
0.50	0.50	0.30	1.0999	1.0923	1.00	1.0004
0.50	0.50	0.40	1.4643	1.4605	1.00	0.9999
0.50	0.50	0.50	1.8408	1.8504	1.00	0.9986

Table 3. Theoretical and experimental values of the scaling law for the cross-covariance for different values of the cross-correlation coefficients in, ρ_a and ρ_b, and between A,B, ρ_{ab}.

Results confirm that α is always equal to one. However, in this case the scaling does not consider the number of aggregated users, but their product, and thus the law is not linear but quadratic. A similar approach was also applied in the particular case in which $\rho_a = \rho_b = \rho_{ab}$, and $\sigma_a = \sigma_b$, that is, when all the consumptions are homogeneous, and with $n_a = n_b$.

Cross-correlation	Coeff		α	
coefficient	theoretical	experimental	theoretical	experimental
0.10	0.40	0.4000	2.00	1.9998
0.20	0.80	0.7914	2.00	1.9997
0.30	1.20	1.2160	2.00	2.0008
0.40	1.60	1.6009	2.00	1.9985
0.50	2.00	2.0059	2.00	1.9992
0.60	2.40	2.3955	2.00	2.0003
0.70	2.80	2.7934	2.00	2.0000
0.80	3.20	3.2043	2.00	1.9999
0.90	3.60	3.6057	2.00	1.9999
0.99	3.96	3.9408	2.00	2.0003

Table 4. Theoretical and experimental values of the scaling law for the cross-covariance between homogeneous groups of consumptions and different values of the cross-correlation coefficient.

Equation 17 then becomes $E[cov_{AB,T}(n_a, n_b)] = n^2 \cdot E[\rho_{ab,\Delta t}] \cdot E[\sigma_{\Delta t}^2]$. The results for different values of the cross-correlation coefficient are described in Table 4. They confirm the theoretical quadratic scaling for cross-covariance.

3.3. Real consumption data: scaling laws for the mean and the variance

The parameters of the scaling laws were also derived for a set of real demand data. The indoor water uses demand series of 82 single-family homes, with a total of 177 inhabitants, in a building belonging to the IIACP (Italian Association of Council Houses) in the town of Latina were considered [29, 30]. The apartments are inhabited by single-income families, belonging to the same low socioeconomic class. The daily demand series of four different days (4 consecutive Mondays) of the 82 users were considered [25]. For each user the different days of consumptions can be considered different realizations of the same stochastic process. In this way the number of customers was artificially extended to about 300, preserving at the same time the homogeneity of the sample. The temporal resolution of each time series is 1 second.

Time	$E[\mu_t]$ (L/min)	$E[\sigma_t^2]$ (L/min)²	a_{var} -
6-7	0.468	1.994	1.2288
7-8	1.066	6.678	1.114
8-9	0.988	7.401	1.0435
9-10	0.891	6.205	1.0756
10-11	0.735	4.336	1.113
11-12	0.791	4.782	1.089
12-13	0.68	4.452	1.092
13-14	0.807	5.322	1.065
14-15	0.827	5.338	1.0688
15-16	0.704	3.857	1.1311
16-17	0.512	2.266	1.1739
17-18	0.634	3.112	1.1666
18-19	0.667	3.594	1.1412
19-20	0.707	5.445	1.0384
20-21	0.68	3.702	1.1253
21-22	0.635	3.412	1.099
22-23	0.397	1.958	1.0771
mean	0.717	4.344	1.1084
confidence limits 95%	0.082	0.759	0.024

Table 5. Estimated parameters of the scaling laws for the experimental data set of Latina (see [25]).

The series were divided into time periods of 1 hour to guarantee the stationarity of the process. In Table 5 the estimated values of the expected values of the mean and the variance of the unit user and the exponent α for the scaling law of the variance are reported. The same exponents for the mean were always trivially equal to 1. In these results the first six hours of the day and the last one were excluded because, during the night hours consumptions are very small and therefore their statistics have a poor significance. It was observed that the mean scales linearly with the number of customers. Differently, the variance shows a slight non-linearity with the number of users. It must be underlined that the average daily value of the exponent α is 1.1, showing that there is a very weak correlation between the considered users.

3.4. Real consumption data: scaling laws for the cross-covariance andcross-correlation coefficient

Considering the consumption signals belonging to homogeneous users, equation 23 is valid and a quadratic scaling law for the cross-covariance should be expected. This behaviour was confirmed by the measured data for all the time intervals considered. In Figure 4 the scaling law of the consumption signals between 11:00 and 12:00 am is graphically reported.

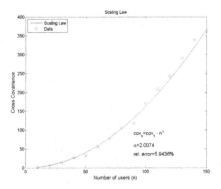

Figure 4. Scaling law for the cross-covariance between 11:00 and 12:00am.

The obtained cross-correlation coefficient between the single user signals was low, being always less than 0.05, but increased noticeably when the number of aggregated users increased, as expected according to equation 22. For groups of 150 aggregated users the cross-correlation coefficient reached the values shown in Table 6. These results enhance the importance of evaluating the cross-correlation degree at different levels of spatial aggregation. Even if the cross-correlation between single-user demand signals is relatively low and less likely to significantly affect the performance of a network, it can largely increase with the spatial aggregation of users, becoming not negligible at those larger scales.

Time	$\rho_{150users}$	Time	$\rho_{150users}$
0-1	0.56	13-14	0.49
1-2	0.6	14-15	0.5
3-4	0.5	15-16	0.42
4-5	0.58	16-17	0.49
5-6	0.36	17-18	0.5
6-7	0.48	18-19	0.39
7-8	0.71	19-20	0.51
8-9	0.61	20-21	0.52
9-10	0.39	21-22	0.39
10-11	0.46	22-23	0.47
11-12	0.57	23-24	0.61
12-13	0.48	-	-

Table 6. Estimated values of the mean cross-correlation coefficients between groups of 150 aggregated user from the experimental data set of Latina.

4. Stochastic simulation of a network

To illustrate the effect of the uncertainty of water demands on the performance of a network, particularly, the effect of the level of correlation between consumptions on the outcome pressure heads, a simple network simulation was performed. The water distribution network of Hanoi (Fujiwara and Khang, 1990) was considered for this matter (Figure 5).

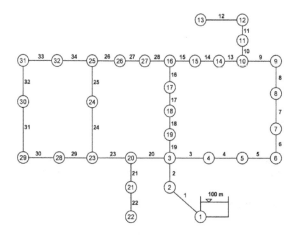

Figure 5. Water distribution network of Hanoi [31].

The data for the Hanoi network were taken from the literature (Fujiwara and Khang, 1990), and the pipe diameters were assumed to be the ones obtained by Cunha and Sousa (2001). The demand data from the literature was used to estimate the number of users at each node, assuming a single-user mean demand of 0.002 l/s. All the users in the network were assumed to be residential and having the same characteristics. The standard deviation of demand was assumed to be 0.06 l/s. The Multivariate Streamflow model [28] was used to generate synthetic stochastic demands with different levels of cross-correlation between the single-users. The nodal demands were then introduced in the network and the performance of the network was simulated using EPANET [32]. For each considered degree of cross-correlation between demands, 100 simulations were performed, resulting in series of pressure heads for each node and for each correlation level.

The first aspect that emerges from the simulations, is that the number of nodes that fail, i.e., which do not satisfy the minimum pressure requirements, increase when the cross-correlation degree increases. Higher correlations imply more synchronous consumptions, leading to pressure failures. Figure 6 illustrates this result.

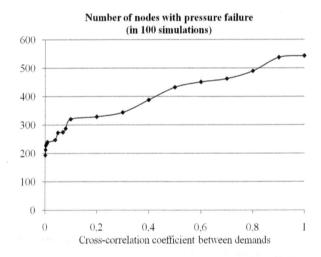

Figure 6. Total number of nodes that do not satisfy minimum pressure requirements in 100 simulations.

Observing Figure 6 it is clear that the cross-correlation between demands significantly affects the outcome pressure heads. The number of nodes that do not satisfy the minimum pressure requirements in the network increase from 194 nodes (total nodes in the network that fail in 100 simulations) when the cross-correlation between demands is equal to 0.001, to a total of 543 nodes when the cross-correlation between demands is 0.999. In other words, the probability of failure increases from 6.3% to 17.5% between the minimum and maximum levels of cross-correlation that were considered.

Another aspect that emerges from the simulations is the increase of standard deviation of the pressure heads at each node of the network, which is illustrated in Figure 7.

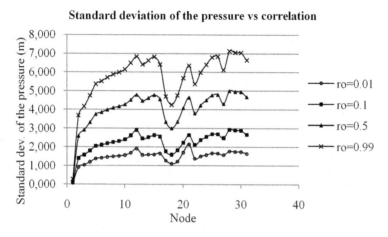

Figure 7. Standard deviation of the pressure heads vs the cross-correlation coefficient between the demands.

The standard deviation of the pressure head verified at each node increases when the cross-correlation between demands increases. The average standard deviation of the pressure heads along the network when the cross-correlation between demands is equal to 0.01 is 1.35m, while the average standard deviation of the pressure heads when the cross-correlation is 0.99, is 5.75m. This means that the cross-correlation increases from 0.01 to 0.99 the standard deviation of the pressure heads increases more than 4 times.

The obtained results clearly show that the level of cross-correlation between demands significantly affects the performance of a network and should, therefore, not be ignored when designing and managing WDS.

5. Conclusions

Understanding and modelling the stochastic nature of water demand represents a challenging field for researchers. Stochastic modelling faces difficulties like scarce availability of data for calibration purposes, high computational efforts associated to simulations, and the complexity of the problem itself. Moreover, the statistical properties of water demand change with the spatial and temporal scales that are used, which makes it even more difficult to accurately model the stochastic structure of demand. The proposed scaling laws represent a step forward in understanding the relation between the parameters that describe probabilistic demands and the spatial and temporal scales in which demands are measured and in which they should be modelled for WDS design or management purposes. The use of scaling laws allow a more accurate quantification of the statistical parameters, like variance and correlation, based on the real demand patterns, number of users at each node and the sampling time that is used. The scaling laws also allow to easily change the scale of the problem, since the statistical parameters and levels of uncertainty can be derived for any desired time or spatial scale.

The scaling laws were derived analytically and validated using synthetically generated stochastic demands and real demand data from Latina, Italy. A good agreement was found between the theoretical expressions, the synthetic demand data and the real demand data. Results show that the mean increases linearly with the number of aggregated users. The variance increases with spatial aggregation according to an exponent that varies between 1 and 2. In theory, for spatially uncorrelated demands the scaling laws is linear and for perfectly correlated demands the scaling law is quadratic. This aspect is clearly verified by the synthetic data. The scaling law for the covariance between 2 groups of users increases according to the product between the numbers of users in each group. The cross-correlation coefficient depends separately on the number of users in each group, and increases towards a limit value. Even for small values of cross-correlation between single-user demands, this parameter cannot be ignored since it significantly increases with the aggregation of consumers.

The performed network simulation considering stochastic demands with different pre-defined levels of correlation show a clear influence of the degree of correlation on the outcome pressure heads: higher levels of correlation lead to larger fluctuations of the pressure heads and to more frequent pressure failures. At this point, the stochastic correlated demands were only used for simulation purposes. However, in future work a similar approach, can be used for design and management purposes. The consideration of correlated stochastic demands will result in more realistic and reliable water distribution networks.

Acknowledgements

The participation of the first author in the study has been supported by Fundação para a Ciência e Tecnologia through Grant SFRH/BD/65842/2009.

Author details

Ina Vertommen[1,2], Roberto Magini[2*], Maria da Conceição Cunha[1] and Roberto Guercio[2]

*Address all correspondence to: roberto.magini@uniroma1.it

1 Department of Civil Engineering, University of Coimbra, Coimbra, Portugal

2 Department of Civil, Building and Environmental Engineering, La Sapienza University of Rome, Rome, Italy

References

[1] Buchberger, S. G., & Wu, L. (1995). Model for instantaneous residential water demands. *Journal of Hydraulic Engineering, ASCE*, 121(3), 232-246.

[2] Buchberger, S. G., & Wells, G. J. (1996). Intensity, Duration, and Frequency of Residential Water Demands. *Journal of Water Resources Planning and Management, ASCE*, 122(1), 11-19.

[3] Garcia, V. J., Garcia-Bartual, R., Cabrera, E., Arregui, F., & Garcia-Serra, J. (2004). Stochastic Model to Evaluate Residential Water Demands. *Water Resources Planning and Management, ASCE*, 130(5), 386-394.

[4] Buchberger, S. G., & Lee, Y.. (1999). Evidence supporting the Poisson pulse hypothesis for residential water demands. *Water Industry systems: Modelling and optimization applications*, 215-227.

[5] Alvisi, S., Franchini, M., & Marinelli, A. (2003). A stochastic model for representing drinking water demand at resindential level. *Water Resources Management*, 17, 197-222.

[6] Blokker, E. J. M., & Vreeburg, J. H. G. (2005). Monte Carlo Simulation of Residential Water Demand: A Stochastic End-Use Model. *Impacts of Global Climate Change, 2005 World Water and Environmental Resources Congress*, Anchorage, Alaska: American Society of Civil Engineers.

[7] Blokker, E. J. M., Vreeburg, J. H. G., & Vogelaar, A. J. (2006). Combining the probabilistic demand model SIMDEUM with a network model Water Distribution System Analysis. *8th Annual Water Distribution Systems Analysis Symposium* , Cincinnati, Ohio, USA.

[8] Blokker, E. J. M., Buchberger, S. G., Vreeburg, J. H. G., & van Dijk, J. C. (2008). Comparison of water demand models: PRP and SIMDEUM applied to Milford, Ohio, data. *WDSA*, Kruger Park, South Africa.

[9] Buchberger, S. G., Blokker, E. J. M., & Vreeburg, J. H. G. (2008). Sizes for Self-Cleaning Pipes in Municipal Water Supply Systems. *WDSA*, Kruger Park, South Africa.

[10] Li, Z., Buchberger, S. G., Boccelli, D., & Filion, Y. (2007). Spatial correlation analysis of stochastic residential water demands. *Water Management Chalenges in Global Change*, London: Taylor & Francis Group.

[11] Blokker, E. J. M., Vreeburg, J. H. G., & van Dijk, J. C. (2010). Simulating Residential Water Demand with a Stochastic End-Use Model. *Journal of Water Resources Planning and Management, ASCE*, 136(1), 19-26.

[12] Moughton, L. J., Buchberger, S. G., Boccelli, D. L., Filion, Y., & Karney, B. W. (2006). *Effect of time step and data aggregation on cross correlation of residential demands*, Cincinnati, Ohio, USA.

[13] Lansey, K. E., & Mays, L. W. (1989). Optimization model for water distribution system design. *Journal of Hydraulic Engineering, ASCE,* 115, 10, 1401-1418.

[14] Xu, C., & Goulter, I. C. (1998). Probabilistic model for water distribution reliability. *Journal of Water Resources Planning and Management, ASCE,* 124(4), 218-228.

[15] Kapelan, Z., Babayan, A. V., Savic, D., Walters, G. A., & Khu, S. T. (2004). Two new approaches for the stochastic least cost design of water distribution systems. *4th World Water Congress: Innovation in Drinking Water Treatmen,* London.

[16] Babayan, A., Savic, D., & Walters, G. (2005). *Multiobjective optimization of water distribution system design under uncertain demand and pipe roughness,* School of Engineering, Computer Science and Mathematics, University of Exeter: Exeter.

[17] Kapelan, Z., Savic, D., & Walters, G. (2005). *An Efficient Sampling-based Approach for the Robust Rehabilitation of Water Distribution Systems under Correlated Nodal Demands,* School of Engineering, Computer Science and Mathematics, University of Exeter: Exeter.

[18] Filion, Y., Adams, B., & Karney, A. (2007). Cross Correlation of Demands in Water Distribution Network Design. *Water Resources Planning and Management, ASCE,* 137-144.

[19] Pasha, K. (2005). Analysis of uncertainty on water distribution hydraulics and water quality. *Impacts of Global Climate Change,* Anchorage, Alaska.

[20] Tessendorff, H. (1972). Problems of peak demands and remedial measures. *Proc., 9th Cong. Int. Water Supply Assoc. int. standing committee on Distribution Problems: subject n. 2.*

[21] Rodriguez-Iturbe, I., Gupta, V. K., & Waymire, E. (1984). Scale considerations in the modeling of temporal rainfall. *Water Resources Research,* 20(11), 1611-1619.

[22] Buchberger, S. G., & Nadimpalli, G. (2004). Leak estimation in water distribution systems by statistical analysis of flow reading. *Water Resources Planning and Management, ASCE* [130, 4], 321-329.

[23] Hutton, C. J., Vamvakeridou-Lyrouda, L. S., Kapelan, Z., & Savic, D. (2011). Uncertainty Quantification and Reduction in Urban Water Systems (UWS) Modelling: Evaluation report.

[24] Giustolisi, O., & Todini, E. (2009). Pipe hydraulic resistance correction in WDN analysis. *Urban Water Journal,* 6(1), 39-52.

[25] Magini, R., Pallavicini, I., & Guercio, R. (2008). Spatial and temporal scaling properties of water demand. *Journal of Water Resources Planning and Management, ASCE,* 134, 276-284.

[26] Rodriguez-Iturbe, I., Marani, M., D'Odorico, P., & Rinaldo, A. (1998). On space-time scaling of cumulated rainfall fields. *Water Resources Research,* 34(12), 3461-3469.

[27] Van Marcke, E. (1983). *Random Fields Analysis and Synthesis (Revised and Expanded New Edition)*, World Scientific Publishing Co. Pte. Ltd.

[28] Fiering, M. B. (1964). Multivariate Technique for Synthetic Hydrology. *Journal of Hydraulics Division - Proceedings of the American Society of Civil Engineers*, 43-60.

[29] Guercio, R., Magini, R., & Pallavicini, I. (2003). Temporal and spatial aggregation in modeling residential water demand. *Water Resources Management II*, UK: WIT Press.

[30] Pallavicini, I., & Magini, R. (2007). Experimental analysis of residential water demand data: probabilistic estimation of peak coefficients at small time scales. *Proc. CCWI2007 & SUWM2007 Conf. Water Management Challenges in Global Change UK*.

[31] Fujiwara, O., & Khang, D. B. (1990). A two phase decomposition method for optimal design of looped water distribution networks. *Water Resources Research*, 26(4), 539-549.

[32] Rossman, L. A. (2000). *Epanet2 Users Manual*, Washington, USA: US EPA.

Chapter 6

Error in Water Meter Measuring Due to Shorter Flow and Consumption Shorter Than the Time the Meter was Calibrated

Dr. Lajos Hovany

Additional information is available at the end of the chapter

1. Introduction

In line with the Measurement Protocol for Water Meters in the Republic of Serbia, a water meter is declared unreliable for water volume measuring at flow rates lower than Q_{min} [1-3]. For a 20 mm rated diameter water meter, used in the households of this country, the minimum flow is about Q_{min}=0.016-0.060 m³/hour. Therefore, water volume is measured unreliably due to leakage at the tap, with a flow of 0.003-0.007 m³/h (drop by drop) and 0.010-0.016 m³/h (jet), at the bathroom tap at 0.010-0.014 m³/h (in a very thin jet), and at the toilet tank at 0.004-0.025 m³/h. As a solution to this problem, the installation of impulse valve, unmeasured-flow reducer, known as UFR, at the water meter is recommended since 2007.

UFR operates based on the difference between the upstream and downstream pressure of 0,4 bar at the valve. For pressure lower than the declared, the UFR closes flow through the water meter, since the spring force is stronger than the force generated by the difference in upstream and downstream pressure. Due to water losses through pipeline leakage, the difference in pressure exceeds the limit value, hence the UFR opens, providing flow at a rate of at least Q_{min} which is then registered by the water meter. UFR manufactured by A.R.I. from Jerusalem is used for adjusting water volume measuring at flow rates lower than 0.026 m³/h.

The papers published so far refer to measurements obtained on individual water meters and on segments of pipelines.

Operation of 33 water meters with UFR was tested in a calibration laboratory in Udine (Italy) [5]. The water meters had a rated diameter of 20 mm, class C, Q_{min}=0.025 m³/h, each 1 to 7

years old. In joint operation of water meters with URF a higher water volume was measured: it was 94% in cases of stagnating water meter propellers, 31.8% for flow rates at the
commencement of the propeller rotation (further designated as Q_a) and 14.4% for Q_{min}. The
valve's most significant contribution was defined for flow at water meter propeller stillstand. Due to the characteristics of water meters installed in the Republic of Serbia, a similar
research is necessary for water meters with 0.025 m^3/h<Q_{min}<0.060 m^3/h.

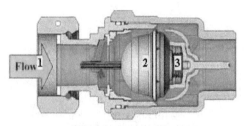

**UFR opens; downstream pressure
equals that of upstream**

Figure 1. Structure of the UFR- manufactured by A.R.I. from Jerusalem (Israel) [4] 1 - flow direction, 2 - UFR plug, 3 -
UFR spring.

Tests were made on parts of the water distribution pipeline in Jerusalem, Larnaca (Cyprus),
on Malta, in Kingston (Tennessee, USA) and in Palermo (Italy). Each testing has been carried
out with two water balancing: one with and another without UFR in operation. By comparing the results of the balancing obtained for the mentioned statuses, the contribution of UFR
to measuring water volume by water meter was determined.

Testing was also implemented in Jerusalem from March 2005 in a duration of 14 months (8
months without UFR and 6 months with UFR) on two systems, where the first comprised
120 and the second 360 water meters [4, 6]. The used class B water meters were with the following characteristics: Q_a=0.012 m^3/h, Q_{min}=0.050 m^3/h and nominal flow Q_n=2.5 m^3/h. With
the usage of UFR, the measured water volume was higher for 16.0-6.1=9.9% (on the system
with 120 water meters) and for 26.0-18.8=7.2% (on the system with 360 water meters).

From October through December 2006, a water supply system with 280, class B and C water
meters, age over 1-15 years was tested in Larnaca with weekly balancing [6]. The water volume measured with UFR was higher for 19.58-9.66=9.92%.

Three tests were made on a system with 26 households on Malta [7-8]. The class D water
meters with rated flow of 1 m^3/h were 5 years old in average. The time interval for water
balancing was one week. The quantity of water measured by UFR on water meters was
more for 26.7-21.2=5.5%, 28-22.2=5.8% and 18.1-12.1=6% than without the use of this valve.

UFRs were installed in a part of a supply network with 35 water meters in Kingston from
June 6 to 10, 2008 [9]. The water meters (aged about 4 years) were calibrated prior to the

measuring. During the operation with URF a higher water consumption was measured: 10.4% for July 2008, 9.5% for August 2008, 4.9% for September 2008, 11.9% for October 2008, 7.6% for November 2008, 8.9% for December 2008, 3.9% for January 2009, 8.4% for February 2009 and 11.6% for March 2009.

Two balancing were made on a part of the supply network with 52 water meters in Palermo: from October 24 through November 14, 2008 without UFR and from December 12, 2008 through January 9, 2009 with UFR [10]. The water meters with 15 mm rated diameter were either 11 years old (33, class C) or older (17 of class B and 2 of class A). During the operation with UFR, the measured volume of consumed water was higher for 28.06-18.91=9.15%.

The use of UFR facilitates the measuring of water consumption at flow rates lower than Q_{min}.

By fulfilling the condition, that the flow of supply network water losses is below 0.026 m³/h and that water consumption exceeds Q_{min}, a water supply network for a single household was set up in the Hydraulic Laboratory of the faculty of Civil Engineering in Subotica. The aim of the test was to confirm the contribution of UFR in measuring water volume by water meter with 20 mm rated diameter and flow of 0.026 m³/h<Q_{min}<0.060 m³/h.

ARI from Jerusalem manufactures UFR with 20 mm rated diameter in three types, designed T10, T20 and T30. Thus, conclusions of testing with valve type T30 could also be controlled by valve types T10 and T20.

Class	Water discharge	Discharged water volume	Permitted error limit		Time between two readings
	m³/h	litres	±%	± litres	minutes
A	Q_{min}=0.06	10	5	0.5	10
	Q_t=0.15	25	2	0.5	10
	Q_n=1.5	100	2	2	4
B	Q_{min}=0.03	5	5	0.25	10
	Q_t=0.12	25	2	0.5	12.5
	Q_n=1.5	100	2	2	4
A	Q_{min}=0.1	20	5	1	12
	Q_t=0.25	30	2	0.6	7.2
	Q_n=2.5	100	2	2	2.4
B	Q_{min}=0.05	20	5	1	24
	Q_t=0.2	25	2	0.5	7.5
	Q_n=2.5	100	2	2	2.4

Table 1. Water volume and calibrated time of water meter of 20 mm rated diameter in the function of typical flows and water meter class.

In line with the Measurement Protocol for Water Meters in the Republic of Serbia, a water meter for water consumption in households is qualified for operation with error below the permitted values, i.e. from ±5% (for Q_{min}) and ±2% (for Q_n and Q_t) from the actual water volume [11]. During calibration, water meter operation errors are checked for the foreseen water volumes.

Through this water volume and discharge, the time for which the meter is calibrated was calculated.

To eliminate the effects of opening and closing the flow switch to measuring errors during calibration, the standard in force in the Republic of Serbia stipulates the following: „The uncertainty introduced into the volume may be considered negligible if the times of motion of the flow switch in each direction are identical within 5% and if this time is less than 1/50 of the total time of the test" [12]. The same recommendations are given by other standards as well [13-14]. Based on that, the following is recommended: „Should there be doubts about whether the operation time of the valve affects the results of the tests, it is recommended that the tests should be made longer, and never under 60 seconds" [15]. That is to say, for neglecting the impact of flow switch manipulation on the water meter's measuring errors the standards offer a solution during the calibration of water meter only.

Water consumption in a single household is implemented by the use of taps, washing machine, dishwasher-machine and shower in the bathroom, likewise the flushing cistern of the toilet and the like. Each consumption is characterised by the opening and closing of flow switch and the duration of water discharge from the pipeline in order to satisfy needs. The duration of consumption in households is shorter than 1 minute in 95% of consumption cases [16]. The error in measuring consumption by water meter, due to manipulating the flow switch, practically manifests as an error due to the duration of consumption shorter than the time the meter was calibrated for [17].

Owing to this fact, the primary aim of this paper is to define measuring errors of consumption shorter than 10 (for Q_{min}), 12.5 (for Q_t) and 4 minutes (for Q_n) of class B water meter with 20 mm rated diameter and flow of Q_n=1.5 m³/h, installed in the water supply pipeline of a single household.

Water meter operation error depends on water meter reading accuracy [18]. The further aim of this paper is to define water consumption measuring error in households shorter than the time the meter was calibrated for, in the function of water meter reading accuracy.

2. The description of the test rig and the tested statuses

In 2010 and 2011, a test rig was set up in the Hydraulic Laboratory of the Faculty of Civil Engineering in Subotica with two-outlets and consumption and water losses were measured by water meters no. 2 and 3.

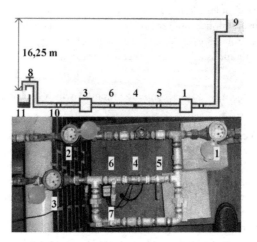

Figure 2. The test rig for water balancing by water maters on water supply system in the Hydraulic Laboratory of the Faculty of Civil Engineering in Subotica 1-3 - water meters, 4 - UFR, 5-8 - shut-off valve, 9 - storage tank, 10 - outlet pipe of water meter no. 3, 11 - balance with a bucket to measure the quantity of water flown through water meter no. 3.

The elements of water balance in the rig were as follows: the volume of the inflow water (measured by water meter no. 1), legal consumption billed and measured (measured by water meter no. 2) and the total volume of water losses, which occurred due to the water meter's inaccurate measuring of flow rates lower than Q_{min} (measured by water meter no. 3).

The operation of the UFR was regulated by shut off valves no. 5, 6 and 7, for example, by shutting down valves no. 5 and 6, the UFR was set out of operation.

Water volumes flown through water meters no. 1, 2 and 3 were defined by the difference of two water meter readings. By measuring time (with stopwatch) between two readings, flows Q_1, Q_2 and Q_3 were calculated by means of the defined water volume. The weight of the water flown through water meter no. 3 was measured by a bucket (16 litres) on a scale. The volume of the water was calculated by the density of the water measured by scale and measuring cylinder.

For the test rig for water balancing in a water supply system, new, calibrated multi-jet propeller water meters with wet mechanism were installed for water temperature of 30°C, and with rated diameter of 20 mm, class B, with the following typical flow rates: Q_a<0.01 m³/h, Q_{min}=0.03 m³/h, Q_t=0.12 m³/h and Q_n=1.5 m³/h.

The used UFR was manufactured by A.R.I. from Jerusalem, with a rated diameter of 20 mm, product type T30. It was installed upstream to water meter no. 3. In line with the manufacturer's recommendation, in order to provide smooth operation of the UFR between water meter no. 3 and shut off valve no. 8, a 6 m long discharge pipe was installed (marked as no. 10 in the attachment).

Water was brought from the reservoir to the rig by gravitation. According to the pressure and flow rate, the rig complied with a single household water supply pipeline.

Two statuses were tested: a) $Q_2 = 0$ and b) $Q_{min} \leq Q_2 \leq Q_n$.

Under the Regulations on the Measurement Protocol for Water Meters of the Republic of Serbia, measuring error of a water meter is defined as:

$$G = \frac{100(Vi - Vc)}{Vc}(\%) \tag{1}$$

where:

Vi - water volume flown through the water meter, registered on the meter's counter, and

Vc - water volume flown through the water meter, measured in the bucket on the scale.

The errors changes in the operation of the water meter for status $Q_2 = 0$ were tested for flow rates $Q_3 < 0.026$ m^3/h for two cases: without UFR and with UFR in operation at water meter no. 3. Applying the criterion, that the error in the water meter's operation is lower than

$$G \leq G_{av} \pm \sigma \ (\%) \tag{2}$$

where:

G_{av} - is the mean error of the water meter in case of steady flow, and

σ - is the standard deviation of the water meter error in steady state flow,

the time needed for getting steady flow, t_{st}, was defined.

After the time required for establishing steady flow t_{st} was determined, the error in water balancing for status $Q_{min} \leq Q_2 \leq Q_n$ and flow rate Q_3 was investigated by the rig:

$$G_b = \frac{100(V_2 + V_3 - V_1)}{V_1}(\%) \tag{3}$$

where:

$V_1 = Q_1 \cdot t_{st}$ - water volume at intake,

$V_2 + V_3 = Q_2 \cdot t_{st} + Q_3 \cdot t_{st}$ - water volume at outlet, and

t_{st} - time for establishing steady flow at water meter no. 3.

The error in balancing was checked for four values of flow rate Q_2 - for two flow rates when $Q_{min} \leq Q_2 \leq Q_t$ and for two flow rates when $Q_t \leq Q_2 \leq Q_n$. The series comprised of 1 to 30 measurements.

Errors in balancing are caused by errors in operation of water meters for measuring water volume: up to ±5% for $Q_{min} \leq (Q_1$ and $Q_2) \leq Q_t$ and up to ±2% for $Q_t \leq (Q_1$ and $Q_2) \leq Q_n$, and undetermined values for $Q_3<Q_{min}$. Since flow rate Q_3 has always been lower than Q_{min}, there is no permitted balancing error limit for the tested statuses.

During accuracy measurements, water meter readings were 2.5 centilitres. Measurement accuracy of water quantity in the vessel was 0.005 kg.

Error in water meter measuring due to consumption shorter than the time the meter was calibrated for: a water supply pipeline was set up in the Hydraulic Laboratory of the Faculty of Civil Engineering in Subotica (in 2011 and 2012), gravitationally supplied from a tank with constant water level, i.e. for 16.25 m higher than the level of the water meter axis. According to both water flow and the water supply pipeline characteristics, the water supply pipeline corresponds to the one of a single household.

Figure 3. Part of the water supply pipeline downstream from the water meter (1) in the Hydraulic Laboratory of the Faculty of Civil Engineering in Subotica. 2 - stop valve, 3 - vessel with scale for measuring water quantity flown through the water meter, 4 - stop-watch, 5 - measuring cylinder and thermometer, 6 - manometer.

During calibration, the reading accuracy of the water meter was 1 decilitre.

A stop valve for starting and stopping water flow was installed at 2.8 m downstream from the water meter.

Water volume flown through the water meter was defined by:

- the difference in reading on the water meter prior and after measuring, and

- measuring water quantity in the vessel (of 15 to 200 litres in volume) and water density.

By measuring time (with stop-watch) between two readings, through the defined water volume in the vessel, the water flow rate Q was calculated.

During accuracy measurements, water meter readings were as follows: 2.5 centilitres, 1 decilitre and 1 litre. Measurement accuracy of water quantity in the vessel was 0.005 kg (for Q_{min}), 0.01 kg (for Q_t) and 0.1 kg (for Q_n).

Error changes in the operation of the water meter described by equation (1), were tested by applying two method by stopping the water meter: according to the valid Protocol of the Republic of Serbia, the status on the water meter and the scale was read prior and after measuring at water meter propeller in stillstand [12].

3. Results

The error changes (G) in the operation of the water meter for status $Q_2=0$ were tested for two cases: without UFR and with UFR in operation at water meter no. 3.

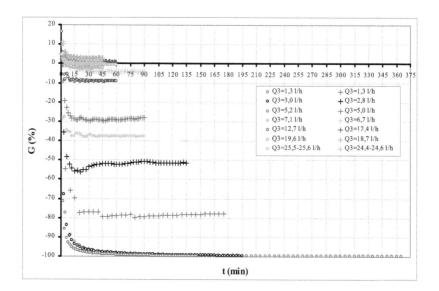

Figure 4. Error changes (G) in the operation of water meter no. 3 during time in the function of Q_3 flow without UFR in operation (circles) and with the UFR in operation (crosses).

With the criterion described by equation (2), steady state flow stabilisation time, t_{st} was defined:

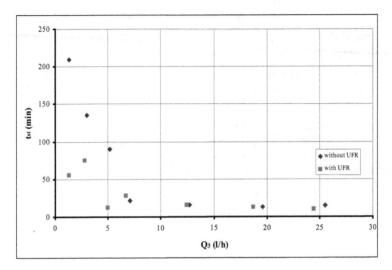

Figure 5. Stabilisation time of water flow in the installation t_{st} in the function of flow Q_3.

Q_3 l/h	t_{st} (min.) without UFR
25.5	15
19.6	13
12.7	16
7.1	22
5.2	90
3	135
1.3	210
Q_3 l/h	t_{st} (min.) with UFR
24.4	10
18.7	13
12.4	16
6.7	28
5	12
2.8	75
1.3	55

Table 2. Stabilisation time of water flow in the installation t_{st} in the function of flow Q_3.

Error in the operation of water meter at steady state flow defined by equation (1) is:

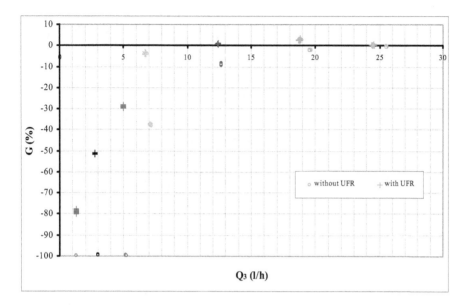

Figure 6. Error (G) in the operation of water meter no. 3 at steady state flow in function of flow rate Q_3, without UFR and with UFR.

The error in the operation of water meter no. 3 without UFR is bigger than with UFR.

The measurements proved that $Q_a < 0.01$ m³/h is between 0.0052 and 0.0067 m³/h.

By increasing flow rate Q_3 towards flow rate Q_a, the influence of UFR operation on measuring water volume increases: for $Q_3=0.0013$ m³/h the contribution is 99.59-78.49=21.1%, and for $Q_3=0.005-0.0052$ m³/h it is 99.5-28.93=70.57%. For flow rates $Q_3 > Q_a$ up to $Q_3=0.026$ m³/h the contribution of UFR to measuring water volume by water meter is decreasing: for flow rate $Q_3=0.0067-0.0071$ m³/h the contribution is 37.67-4.07=33.6%, and for $Q_3=0.0244-0.0255$ m³/h it is 0.47+0.36=0.83%.

Without the UFR in operation, the water meter always shows lower water volume than the real value. With the UFR in operation, at flow rate $Q_3=0.01$ m³/h, the operation of the water meter is changing: for flow rate $Q_3 < 0.01$ m³/h the water meter shows lower volume than the real value, while for $Q_3 > 0.01$ m³/h it shows higher value (not exceeding 2.7%) than the real one.

For status $Q_{min} \leq Q_2 \leq Q_n$ balancing errors described by equation (3) are:

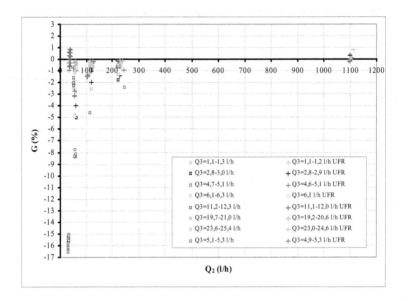

Figure 7. Error (G_b) in water balancing on the rig in the function of flow rates Q_2 and Q_3, without UFR (left) and with UFR (right) in operation.

The highest balancing error values for the tested statuses are +0.9 and -16.6%. By increasing flow rate Q_2 for both statuses (without UFR and with UFR at water meter no. 3) errors in balancing decrease.

With UFR in operation, the most significant contribution in measuring water volume by water meter (15.7-0=15.7%, mean value of 30 measurments) was determined at operations at flow rates Q_2=0.03-0.0385 m³/h and Q_3=0.0049-0.0053 m³/h.

At steady flow slightly lesser than Q_a, UFR efficiency increases from T10 to T30: water meter equipped with UFR type T10 measures for 99.53-84.14=15.39% more water than in status without UFR, while the increase with type T20 is 99.33-68.55=30.78% and with type T30 it is 99.5-28.93=70.57%.

This conclusion was verified for a single household water supply network by the results of five measurements for each status: water meter equipped with type T10 UFR measured (at flow rates Q_2=51.3-51.4 l/h and Q_3=5.2-5.6 l/h) maximum 9.89-6.84=3.05% more water consumption of than without UFR, with type T20 (at flow rate of Q_2=56.9-60 l/h and Q_3=5.3-5.5 l/h) the most significant improvement was 9.06-4.19=4.87%, while with type T30 (at flow rates of Q_2=55.9-58.5 l/h and Q_3=4.9-5.1 l/h) the result is 8.4-0.97=7.43%.

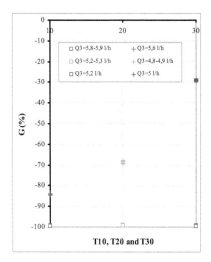

Figure 8. Error (G) in the operation of water meter no. 3 at steady state flow in function of flow rate $Q_3=Q_a$ without UFR (left) and with UFR (right) type T10, T20 and T30.

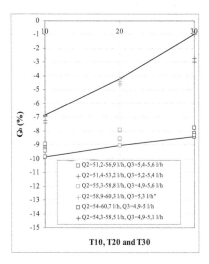

Figure 9. Water balance error (G_b) of the tested water supply network without UFR type T10, T20 and T30 (quadrat) and equipped with one of the UFR type (cross).

Error in water meter measuring due to consumption shorter than the time the meter was calibrated for: each measurement was repeated 5-30 times.

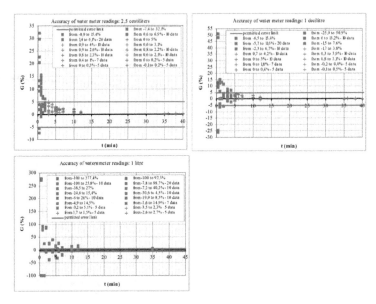

Figure 10. Error in water meter measuring at Q_{min} flow during water flow shorter than the time for which the meter was calibrated (10 min.) in the function of water meter reading accuracy.

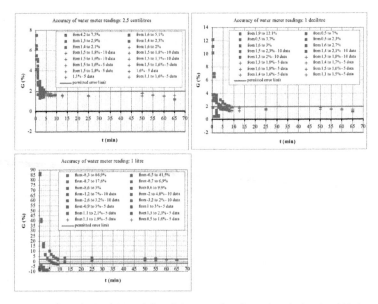

Figure 11. Error in water meter measuring at Q_t flow during water flow shorter than the time for which the meter was calibrated (12.5 min.) in the function of water meter reading accuracy.

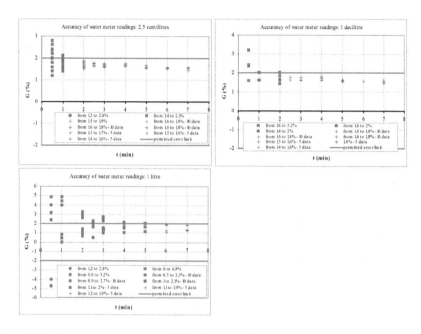

Figure 12. Error in water meter measuring at Q_n flow during water flow shorter than the time for which the meter was calibrated (4 min.) in the function of water meter reading accuracy.

The minimum time at which measuring errors undoubtedly are within the permitted limit is defined for the calibrated flows of the water meter as follows:

Reading accuracy	Dis- charge	Minimum time (minutes)
2.5 centilitres	Q_{min}	3
	Q_t	5
	Q_n	2
1 decilitre	Q_{min}	5
	Q_t	9
	Q_n	2.5
1 litre	Q_{min}	36
	Q_t	60
	Q_n	6

Table 3. Minimum calibrated flow duration time in the function of reading accuracy.

4. Discussion

Measurement results confirm and specify the conclusions of tests made in Udine in relation to the contribution of UFR to water meter operation: a) that the most significant contribution is at flow, when water meter propeller is steady (specifically for Q_a), and b) that this contribution is decreasing from flow Q_a towards flow Q_{min}.

The same conclusion is valid for test results on the UFR contribution at parts of the water supply network. Providing similar water consumption during testings on the test rig without UFR and with UFR enabled specifying the results.

In the case, when consumption time was shorter than the time the meter was calibrated for, the range of the meter measuring error exceeded the range of permitted errors.

Reading Accuracy	Dis- charge	Error range (%)	
		During experiment	During calibration
2.5 centilitar	Q_{min}	from -7.4 to 32.1	
	Q_t	from 4.2 to 7.5	
	Q_n	from 1.2 to 2.8	
1 decilitar	Q_{min}	from -25.9 to 50.9	±5
	Q_t	from 1.9 to 12.1	±2
	Q_n	from 1.6 to 3.2	±2
1 litar	Q_{min}	from -100 to 277.4	
	Q_t	from -8.3 to 86.9	
	Q_n	from -4.8 to 4.8	

Table 4. Error ranges of water meter measuring for calibrated discharges in the function of water meter reading accuracy (class B, rated diameter of 20 mm and discharge of Q_n=1.5 m³/h), for a duration of 0.5 minutes flow.

It means that during consumption shorter than the time the water meter was calibrated for, measuring by the meter is unreliable in 95% of the consumption shorter than 1 minute. During a discharge of 0.5 minutes, the error may be even 277.4%.

In order to improve water consumption measuring in households, it is necessary to provide conditions for measuring consumption at flow lower than Q_{min} and duration shorter than the time the meter has been calibrated for. Such conditions may be created in supply pipelines with water storage tanks in households. Only such systems are appropriate in which all the water needed in a household flows through this storage tank [19].

The UFR should be installed on the outlet pipe from the tank to the household. Signalising the start of the UFR's operation may initiate works on eliminating water losses due to leakage on the tap, bathroom battery and the flushing cistern. This way, the use of UFR would serve to protect the interest of the households.

Recently, it has been a frequent practice, that a water meter is installed on the inlet pipe of the storage tank near the building [10, 20-25]. Such water meters provide the opportunity to ensure a minimum period of time, at which water meter measurement error surely stays within the permitted error limit.

By solving water quality problems in these storage tanks, the above mentioned conditions will, first and foremost, be provided in settlements already having such water supply networks in place, e.g. in settlements without continuous potable water supply (e.g. in Mozambique, Yemen, Jordanian, Lebanon, Palestine, on the Mediterranean in Europe) and in water supply networks designed in the XIXth century (e.g. the UK) [21, 23, 26-30].

5. Conclusion

According to pressure and water flow rate, the test rig in the Hydraulic Laboratory of the Faculty of Civil Engineering Subotica presented a case of a single household water pipeline. The water meters are produced by Potiski vodovodi Ltd. from Horgos with rated diameter of 20 mm, class B, and the following flow rate characteristics: the flow at starting the water meter propeller (Q_a) is between 0.0052 and 0.0067 m^3/h, Q_{min}=0.03 m^3/h, Q_t=0.12 m^3/h and Q_n=1.5 m^3/h. Upstream from water meter no. 3 an UFR was installed, manufactured by A.R.I. from Jerusalem, with rated diameter of 20 mm, product type T30.

During accuracy measurements, water meter readings were 2.5 centilitres. Measurement accuracy of water quantity in the vessel was 0.005 kg.

According to the measurement results, the most significant contribution of the UFR in measuring water volume by a water meter of a single household takes place at water losses by flow rate Q_a, prior to starting the water meter propeller:

a) 70.57% at flow rates Q_2=0 and Q_3=0.005-0.0052 m^3/h, and

b) 15.7% at flow rates Q_2=0.03-0.0385 m^3/h and Q_3=0.0049-0.0053 m^3/h.

The effects of UFR types T10 and T20 manufactured by ARI from Jerusalem on water measuring accuracy are less than that of type T30.

For calibration discharges, foreseen by the Protocol, through the testing of class B flow meter with 20 mm rated diameter and discharge of Q_n=1,5 m^3/h, it has been established, that:

- in the case, when consumption time was shorter than the time the meter was calibrated for, at discharges of , the range of the meter measuring error exceeded the range of permitted errors,

- the biggest errors occur at Q_{min} - for example, at discharge lasting for 0.5 minute, the error may be even 32.1% (for water meter reading accuracy of 2.5 centilitres), or even 50.9% (for water meter reading accuracy of 1 decilitre), or even 277.4% (for water meter reading accuracy of 1 litre), and

- to provide that measuring errors of calibrated flows be lower than the permitted ones, the minimum time for measuring these flows is 5 minutes (for water meter reading accuracy of 2.5 centilitres), 9 minutes (for water meter reading accuracy of 1 decilitre), or 60 minutes (for water meter reading accuracy of 1 litre).

Since it concerns 95% of water consumption measurement, such testings are necessary for all types of water meters used in the supply networks of this country.

Author details

Dr. Lajos Hovany*

Address all correspondence to: hovanyl@gf.uns.ac.rs

Faculty of Civil Engineering Subotica, Republic of Serbia

References

[1] Hovanj, L. (2010). Minimum Time Period Between Reading off Flow Meters (Minimalno vreme između dva očitavanja vodomera). *Zbornik radova Građevinskog fakulteta u Subotici*, 19(1), 105-113.

[2] Hovány, L. (2011). UFR Suppored Water Meter (Impulzust keltő szeleppel kiegészített vízmérő). *Hidrológiai Közlöny*, 91(2), 23-26.

[3] Hovány, L. (2011). The Contribution of UFR in Measuring Water Volume by Water Meter in a Single Household. *In: EXPRES 2011. 3rd IEEE International Symposium on Exploitation of Renewable Energy Sources*, Procedings, March 11-12, Subotica, Serbia, 75-78.

[4] Davidesko, A. (2007). UFR- an innovative solution for water meter under registration. *In: Proceeding of the Water Loss 2007 Conference*, September 23-26. Bucharest, Romania, 704-709, http://waterloss2007.com/pdf_vortraege/Mittwoch/B9-3.pdf.

[5] Fantozzi, M. Reduction of customer meter under-registration by optimal economic replacement based on meter accuracy testing programme and Unmeasured Flow Reducers.   http://www.studiomarcofantozzi.it/Aprile09/Fantozzi%20AppLosses %20WL09%20Paper%20V11%20250209.pdf.

[6] Yaniv, S. (2009). Reduction of Apparent Losses Using the UFR (Unmeasured-Flow Reducer). Case Studies. *In: Efficient 2009 Conference*, October 25-29, Sydney, Australia, 1-8,   http://www.arivalves.com/PDF/UFR/Reductions%20of%20Apparent%20Losses.pdf.

[7] Rizzo, A., Bonello, M., & Galea St. John, S. (2007). Trials to Quantify and Reduce in-situ Meter Under-Registration. *In: Proceeding of the Water Loss 2007 Conference*, September 23-26, Bucharest, Romania, 695-703, http://www.arivalves.com/PDF/UFR/Trials%20to%20Quantify%20and%20Reduce%20in-situ%20Meter%20Under-.pdf.

[8] Rizzo, A., Vermersch, M., Galea St. John, S., Micallef, G., Riolo, S., & Pace, R. (2007). Apparent Water Loss Control. *The way Forward*, http://www.iwaom.org/datosbda/Descargas/48.pdf.

[9] Haas, B., & Barger, W. (2009). Wattsbar utility UFR pilot summary. 04.03.2009., http://www.aymcdonald.com/IntranetSales/ufr/WATTSBAR%204_3_09.pdf.

[10] Fantozzi, M., Criminisi, A., Fontanazza, C. M., Freni, G., & Lambert, A. Investigations into under-registration of customer meters in Palermo (Italy) and the effect of introducing Unmeasured Flow Reducers. http://www.arivalves.com/PDF/UFR/Fantozzi%20et%20Al%20Palermo%20case%20study%20V4%20030309.pdf.

[11] Pravilnik o metrološkim uslovima za vodomere. (1986). *Službeni list SFRJ* [51], 1509-1513.

[12] SRPS EN 14154-3: (2010). (en) Water meters (Merila protoka vode). Part 3: Test methods and equipment (Deo 3: Metode ispitivanja i oprema).

[13] ISO 4064-3:2005(E) Measurement of water flow in fully charged closed conduits- Meters for cold portable water and hot water. Part 3: Test methods and equipment.

[14] OIML R 49-2: 2006 (E) Water meters intended for the metering of cold potable water and hot water. Part 2: Test methods.

[15] Arregui, F., Cabrera Jr, E., & Cobacho, R. (2006). Integrated Water Meter Management. London: IWA Publishing.

[16] Buchberger, S. G., & Wells, G. J. (1996). Intensity, Duration, and Frequency of Residential Water Demands. *Journal of Water Resources Planning and Management*, 122(1), 11-19.

[17] Hovány, L. (2012). Error in Water Meter Measuring at Water Flow Rate Exceeding Q_{min} . *In: EXPRES 2012. 4rd IEEE International Symposium on Exploitation of Renewable Energy Sources*, Final Program, March 09-10, Subotica, Serbia, 63-65.

[18] Hovanj, L. (2010). Error in Water Balancing at Discharges Lower than Q_{min}, Measured by Water Meter (Greška bilansiranja vode vodovoda pri manjem proticaju od Q_{min}-merena vodomerom). *Voda i sanitarna tehnika*, 40(3), 35-42.

[19] Charalambous, B., Charalambous, S., & Ioannou, I. Meter Under-Registration caused by Ball Valves in Roff Tanks. 710-719, http://www.aymcdonald.com/IntranetSales/ufr/Meter%20Under-Registration%20caused%20by%20Ball%20Valves%20in%20Roof%20(2).pdf.

[20] Rizzo, A., & Cilia, J. (2005). Quantifying Meter Under-Registration Caused by the Ball Valves of Roof Tanks (for Indirect Plumbing Systems). *In: IWA Specialised Conference Leakage 2005, Conference Proceedings*, Halifax, Nova Scotia, Canada, p.106, 1-12.

[21] Cobacho, R., Arregui, F., Cabrera, E., & Cabrera Jr, E. (2007). Private Water Storage Tanks: Evaluating Their Inefficiencies. *In: Efficient 2007. The 4th IWA Specialised Conference of Efficient Use od Urban Water Supply, Conference Procedings*, Jeju Island, Korea, (1-8), http://www.ita.upv.es/idi/descargaarticulo.php?id=173.

[22] Arregui, F. J., Pardo, M. A., Parra, J. C., & Soriano, J. (2007). Quantification of meter errors of domestic users: a case study. *In: Proceedings of the Water Loss 2007 Conference*, Bucharest, September 23-26, 1-11, http://www.ita.upv.es/idi/descargaarticulo.php?id=200.

[23] Criminisi, A., Fontanazza, C. M., Freni, G., & La Loggia, G. (2009). Evaluation of the apparent losses caused by water meter under-registration in intermittent water supply. *Water Science & Technology-WST*, 60(9), 2373-2382.

[24] De Marchis, M., Freni, G., & Napoli, E. A Numerical Unsteady Friction Model for the Transient Flow Arising During the Filling Process of Intermittent Water Distribution Systems. 1-6, ftp://ftp.optimale.com.br/CCWI2011/papers/253.pdf.

[25] Fontanazza, C. M., Freni, G., La Loggia, G., Notaro, V., & Puleo, V. A Performance-Based Tool for Prioritising Water Meter Substitution in a Urban Distribution Network. 1-7, ftp://ftp.optimale.com.br/CCWI2011/papers/191.pdf.

[26] Evison, L., & Sunna, N. (2001). Microbial Regrowth in Household Water Storage Tanks. *Journal American Water Works Association*, 93(9), 85-94.

[27] Coelho, S. T., James, S., Sunna, N., Abu Jaish, A., & Chatila, J. (2003). Controlling water quality in intermittent supply systems. *Water Science &.Techology- Water Supply*, 3(1-2), 119-125.

[28] Trifunović, N. (2006). Introduction to Urban Water Distribution. London, Leiden, New York, Philadelphia, Singapore: Taylor&Francis.

[29] Matsinhe, N. P., Juizo, J., & Persson, K. M. (2008). The effect of intermittent supply and household storage on the quality of drinking water in Maputo. *In: Challenges and Opportunities for Safe Water Supply in Mozambique*, http://lup.lub.lu.se/luur/download?func=downloadFile&recordOId=1227316&fileOId=1227505.

[30] Trifunovic, N. (2012). Water Transport & Distribution. Delft: UNESCO-IHE 05.03.2012., oc.its.ac.id/ambilfile.php?idp=1897.

Permissions

The contributors of this book come from diverse backgrounds, making this book a truly international effort. This book will bring forth new frontiers with its revolutionizing research information and detailed analysis of the nascent developments around the world.

We would like to thank Avi Ostfeld, for lending his expertise to make the book truly unique. He has played a crucial role in the development of this book. Without his invaluable contribution this book wouldn't have been possible. He has made vital efforts to compile up to date information on the varied aspects of this subject to make this book a valuable addition to the collection of many professionals and students.

This book was conceptualized with the vision of imparting up-to-date information and advanced data in this field. To ensure the same, a matchless editorial board was set up. Every individual on the board went through rigorous rounds of assessment to prove their worth. After which they invested a large part of their time researching and compiling the most relevant data for our readers. Conferences and sessions were held from time to time between the editorial board and the contributing authors to present the data in the most comprehensible form. The editorial team has worked tirelessly to provide valuable and valid information to help people across the globe.

Every chapter published in this book has been scrutinized by our experts. Their significance has been extensively debated. The topics covered herein carry significant findings which will fuel the growth of the discipline. They may even be implemented as practical applications or may be referred to as a beginning point for another development. Chapters in this book were first published by InTech; hereby published with permission under the Creative Commons Attribution License or equivalent.

The editorial board has been involved in producing this book since its inception. They have spent rigorous hours researching and exploring the diverse topics which have resulted in the successful publishing of this book. They have passed on their knowledge of decades through this book. To expedite this challenging task, the publisher supported the team at every step. A small team of assistant editors was also appointed to further simplify the editing procedure and attain best results for the readers.

Our editorial team has been hand-picked from every corner of the world. Their multi-ethnicity adds dynamic inputs to the discussions which result in innovative outcomes. These outcomes are then further discussed with the researchers and contributors who give their valuable feedback and opinion regarding the same. The feedback is then collaborated with the researches and they are edited in a comprehensive manner to aid the understanding of the subject.

Apart from the editorial board, the designing team has also invested a significant amount of their time in understanding the subject and creating the most relevant covers. They scrutinized every image to scout for the most suitable representation of the subject and create an appropriate cover for the book.

The publishing team has been involved in this book since its early stages. They were actively engaged in every process, be it collecting the data, connecting with the contributors or procuring relevant information. The team has been an ardent support to the editorial, designing and production team. Their endless efforts to recruit the best for this project, has resulted in the accomplishment of this book. They are a veteran in the field of academics and their pool of knowledge is as vast as their experience in printing. Their expertise and guidance has proved useful at every step. Their uncompromising quality standards have made this book an exceptional effort. Their encouragement from time to time has been an inspiration for everyone.

The publisher and the editorial board hope that this book will prove to be a valuable piece of knowledge for researchers, students, practitioners and scholars across the globe.

List of Contributors

Helena Alegre and Sérgio T. Coelho
LNEC - National Civil Engineering Laboratory, Lisbon,, Portugal

Ivo Pothof
Deltares, MH Delft, The Netherlands
Delft University of Technology, Department of Water Management, Stevinweg, CN Delft, The Netherlands

Bryan Karney
University of Toronto, Canada and HydraTek and Associates Inc., Canada

Thomas Bernard and Oliver Krol,
Business Department MRD, Fraunhofer Institute of Optronics, System Technologies and Image Exploitation IOSB, Karlsruhe, Germany

Thomas Rauschenbach and Divas Karimanzira
Fraunhofer Application Center System Technology AST, Fraunhofer Institute of Optronics, System Technologies and Image Exploitation IOSB, Ilmenau, Germany

H. M. Ramos
Instituto Superior Técnico, Technical University of Lisbon, Portugal

L. H. M. Costa
Federal University of Ceará, Brazil

F. V. Gonçalves
Federal University of Mato Grosso do Sul, Brazil

Ina Vertommen
Department of Civil Engineering, University of Coimbra, Coimbra, Portugal
Department of Civil, Building and Environmental Engineering, La Sapienza University of Rome, Rome, Italy

Roberto Magini and Roberto Guercio
Department of Civil, Building and Environmental Engineering, La Sapienza University of Rome, Rome, Italy

Maria da Conceição Cunha
Department of Civil, Building and Environmental Engineering, La Sapienza University of Rome, Rome, Italy

Dr. Lajos Hovany
Faculty of Civil Engineering Subotica, Republic of Serbia

Printed in the USA
CPSIA information can be obtained
at www.ICGtesting.com
JSHW011334221024
72173JS00003B/151

9 781632 396211